T0138082

Lecture Notes in Computer Science 14613

Founding Editors

Gerhard Goos
Juris Hartmanis

The series Lecture Notes in Computer Science (LNCS), including its subseries Lecture Notes in Artificial Intelligence (LNAI) and Lecture Notes in Bioinformatics (LNBI), has established itself as a medium for the publication of new developments in computer science and information technology research, teaching, and education.

LNCS enjoys close cooperation with the computer science R & D community, the series counts many renowned academics among its volume editors and paper authors, and collaborates with prestigious societies. Its mission is to serve this international community by providing an invaluable service, mainly focused on the publication of conference and workshop proceedings and postproceedings. LNCS commenced publication in 1973.

Nazli Goharian · Nicola Tonellotto · Yulan He ·
Aldo Lipani · Graham McDonald ·
Craig Macdonald · Iadh Ounis
Editors

Advances in Information Retrieval

46th European Conference on Information Retrieval, ECIR 2024
Glasgow, UK, March 24–28, 2024
Proceedings, Part VI

Editors
Nazli Goharian
Georgetown University
Washington, WA, USA

Yulan He 🅞
King's College London
London, UK

Graham McDonald 🅞
University of Glasgow
Glasgow, UK

Iadh Ounis 🅞
University of Glasgow
Glasgow, UK

Nicola Tonellotto 🅞
University of Pisa
Pisa, Italy

Aldo Lipani 🅞
University College London
London, UK

Craig Macdonald 🅞
University of Glasgow
Glasgow, UK

ISSN 0302-9743   ISSN 1611-3349 (electronic)
Lecture Notes in Computer Science
ISBN 978-3-031-56071-2   ISBN 978-3-031-56072-9 (eBook)
https://doi.org/10.1007/978-3-031-56072-9

This Springer imprint is published by the registered company Springer Nature Switzerland AG
The registered company address is: Gewerbestrasse 11, 6330 Cham, Switzerland

Paper in this product is recyclable.

Preface

The 46th European Conference on Information Retrieval (ECIR 2024) was held in Glasgow, Scotland, UK, during March 24–28, 2024, and brought together hundreds of researchers from the UK, Europe and abroad. The conference was organised by the University of Glasgow, in cooperation with the British Computer Society's Information Retrieval Specialist Group (BCS IRSG) and with assistance from the Glasgow Convention Bureau.

These proceedings contain the papers related to the presentations, workshops, tutorials, doctoral consortium and other satellite tracks that took place during the conference. This year's ECIR program boasted a variety of novel work from contributors from all around the world. In addition, we introduced a number of novelties in this year's ECIR. First, ECIR 2024 included for the first time a new "Findings" track, which was offered to some full papers that were deemed to be solid, but which could not make the main conference track. Second, ECIR 2024 ran a new special IR4Good track that presented high-quality, high-impact, original IR-related research on societal issues (such as algorithmic bias and fairness, privacy, and transparency) at the interdisciplinary level (e.g., philosophy, law, sociology, civil society), which go beyond the purely technical perspective. Third, ECIR 2024 featured a new innovation called the "Collab-a-thon", intended to provide an opportunity for participants to foster new collaborations that could lead to exciting new research, and forge lasting relationships with like-minded researchers. Finally, ECIR 2024 introduced a new award to encourage and recognise researchers who have made significant contributions in using theory to develop the information retrieval field. The award was named after Professor Cornelis "Keith" van Rijsbergen (University of Glasgow), a pioneer in modern information retrieval, and a strong advocate of the development of models and theories in information retrieval.

The ECIR 2024 program featured a total of 578 papers from authors in 61 countries in its various tracks. The final program included 57 full papers (23% acceptance rate), an additional 18 finding papers, 36 short papers (24% acceptance rate), 26 IR4Good papers (41%), 18 demonstration papers (56% acceptance rate), 9 reproducibility papers (39% acceptance rate), 8 doctoral consortium papers (57% acceptance rate), and 15 invited CLEF papers. All submissions were peer-reviewed by at least three international Program Committee members to ensure that only submissions of the highest relevance and quality were included in the final ECIR 2024 program. The acceptance decisions were further informed by discussions among the reviewers for each submitted paper, led by a Senior Program Committee member. Each track had a final PC meeting where final recommendations were discussed and made, trying to reach a fair and equal outcome for all submissions.

The accepted papers cover the state-of-the-art in information retrieval and recommender systems: user aspects, system and foundational aspects, artificial intelligence & machine learning, applications, evaluation, new social and technical challenges, and other topics of direct or indirect relevance to search and recommendation. As in previous years, the ECIR 2024 program contained a high proportion of papers with students as first authors, as well as papers from a variety of universities, research institutes, and commercial organisations.

In addition to the papers, the program also included 4 keynotes, 7 tutorials, 10 workshops, a doctoral consortium, an IR4Good event, a Collab-a-thon and an industry day. Keynote talks were given by Charles L. A. Clarke (University of Waterloo), Josiane Mothe (Université de Toulouse), Carlos Castillo (Universitat Pompeu Fabra), and this year's Keith van Rijsbergen Award winner, Maarten de Rijke (University of Amsterdam). The tutorials covered a range of topics including explainable recommender systems, sequential recommendation, social good applications, quantum for IR, generative IR, query performance prediction and PhD advice. The workshops brought together participants to discuss narrative extraction (Text2Story), knowledge-enhanced retrieval (KEIR), online misinformation (ROMCIR), understudied users (IR4U2), graph-based IR (IRonGraphs), open web search (WOWS), technology-assisted review (ALTARS), geographic information extraction (GeoExT), bibliometrics (BIR) and search futures (SearchFutures).

The success of ECIR 2024 would not have been possible without all the help from the strong team of volunteers and reviewers. We wish to thank all the reviewers and meta-reviewers who helped to ensure the high quality of the program. We also wish to thank: the reproducibility track chairs Claudia Hauff and Hamed Zamani, the IR4Good track chairs Ludovico Boratto and Mirko Marras, the demo track chairs Giorgio Maria Di Nunzio and Chiara Renso, the industry day chairs Olivier Jeunen and Isabelle Moulinier, the doctoral consortium chairs Yashar Moshfeghi and Gabriella Pasi, the CLEF Labs chair Jake Lever, the workshop chairs Elisabeth Lex, Maria Maistro and Martin Potthast, the tutorial chairs Mohammad Aliannejadi and Johanne R. Trippas, the Collab-a-thon chair Sean MacAvaney, the best paper awards committee chair Raffaele Perego, the sponsorship chairs Dyaa Albakour and Eugene Kharitonov, the proceeding chairs Debasis Ganguly and Richard McCreadie, and the local organisation chairs Zaiqiao Meng and Hitarth Narvala. We would also like to thank all the student volunteers who worked hard to ensure an excellent and memorable experience for participants and attendees. ECIR 2024 was sponsored by a range of learned societies, research institutes and companies. We thank them all for their support. Finally, we wish to thank all of the authors and contributors to the conference.

March 2024

Nazli Goharian
Nicola Tonellotto
Yulan He
Aldo Lipani
Graham McDonald
Craig Macdonald
Iadh Ounis

Organization

General Chairs

Craig Macdonald   University of Glasgow, UK
Graham McDonald   University of Glasgow, UK
Iadh Ounis   University of Glasgow, UK

Program Chairs – Full Papers

Nazli Goharian   Georgetown University, USA
Nicola Tonellotto   University of Pisa, Italy

Program Chairs – Short Papers

Yulan He   King's College London, UK
Aldo Lipani   University College London, UK

Reproducibility Track Chairs

Claudia Hauff   Spotify & TU Delft, Netherlands
Hamed Zamani   University of Massachusetts Amherst, USA

IR4Good Chairs

Ludovico Boratto   University of Cagliari, Italy
Mirko Marras   University of Cagliari, Italy

Demo Chairs

Giorgio Maria Di Nunzio Università degli Studi di Padova, Italy
Chiara Renso   ISTI - CNR, Italy

Industry Day Chairs

Olivier Jeunen   ShareChat, UK
Isabelle Moulinier   Thomson Reuters, USA

Doctoral Consortium Chairs

Yashar Moshfeghi   University of Strathclyde, UK
Gabriella Pasi   Università degli Studi di Milano Bicocca, Italy

CLEF Labs Chair

Jake Lever   University of Glasgow, UK

Workshop Chairs

Elisabeth Lex   Graz University of Technology, Austria
Maria Maistro   University of Copenhagen, Denmark
Martin Potthast   Leipzig University, Germany

Tutorial Chairs

Mohammad Aliannejadi   University of Amsterdam, Netherlands
Johanne R. Trippas   RMIT University, Australia

Collab-a-thon Chair

Sean MacAvaney   University of Glasgow, UK

Best Paper Awards Committee Chair

Raffaele Perego   ISTI-CNR, Italy

Sponsorship Chairs

Dyaa Albakour   Signal AI, UK
Eugene Kharitonov   Google, France

Proceeding Chairs

Debasis Ganguly   University of Glasgow, UK
Richard McCreadie   University of Glasgow, UK

Local Organisation Chairs

Zaiqiao Meng   University of Glasgow, UK
Hitarth Narvala   University of Glasgow, UK

Senior Program Committee

Mohammad Aliannejadi   University of Amsterdam, Netherlands
Omar Alonso   Amazon, USA
Giambattista Amati   Fondazione Ugo Bordoni, Italy
Ioannis Arapakis   Telefonica Research, Spain
Jaime Arguello   The University of North Carolina at Chapel Hill,
                               USA
Javed Aslam   Northeastern University, USA
Krisztian Balog   University of Stavanger & Google Research,
                               Norway
Patrice Bellot   Aix-Marseille Université CNRS (LSIS), France
Michael Bendersky   Google, USA
Mohand Boughanem   IRIT University Paul Sabatier Toulouse, France
Jamie Callan   Carnegie Mellon University, USA
Charles Clarke   University of Waterloo, Canada
Fabio Crestani   Università della Svizzera italiana (USI),
                               Switzerland
Bruce Croft   University of Massachusetts Amherst, USA
Maarten de Rijke   University of Amsterdam, Netherlands
Arjen de Vries   Radboud University, Netherlands
Tommaso Di Noia   Politecnico di Bari, Italy
Carsten Eickhoff   University of Tübingen, Germany
Tamer Elsayed   Qatar University, Qatar

Benjamin Piwowarski   CNRS/ISIR/Sorbonne Université, France
Paolo Rosso   Universitat Politècnica de València, Spain
Mark Sanderson   RMIT University, Australia
Philipp Schaer   TH Köln (University of Applied Sciences),
                          Germany
Ralf Schenkel   Trier University, Germany
Christin Seifert   University of Marburg, Germany
Gianmaria Silvello   University of Padua, Italy
Fabrizio Silvestri   University of Rome, Italy
Mark Smucker   University of Waterloo, Canada
Laure Soulier   Sorbonne Université-ISIR, France
Torsten Suel   New York University, USA
Hussein Suleman   University of Cape Town, South Africa
Paul Thomas   Microsoft, USA
Theodora Tsikrika   Information Technologies Institute/CERTH,
                          Greece
Suzan Verberne   LIACS/Leiden University, Netherlands
Marcel Worring   University of Amsterdam, Netherlands
Andrew Yates   University of Amsterdam, Netherlands
Shuo Zhang   Bloomberg, UK
Min Zhang   Tsinghua University, China
Guido Zuccon   The University of Queensland, Australia

Program Committee

Amin Abolghasemi   Leiden University, Netherlands
Sharon Adar   Amazon, USA
Shilpi Agrawal   Linkedin, USA
Mohammad Aliannejadi   University of Amsterdam, Netherlands
Satya Almasian   Heidelberg University, Germany
Giuseppe Amato   ISTI-CNR, Italy
Linda Andersson   Artificial Researcher IT GmbH TU Wien, Austria
Negar Arabzadeh   University of Waterloo, Canada
Marcelo Armentano   ISISTAN (CONICET - UNCPBA), Argentina
Arian Askari   Leiden University, Netherlands
Maurizio Atzori   University of Cagliari, Italy
Sandeep Avula   Amazon, USA
Hosein Azarbonyad   Elsevier, Netherlands
Leif Azzopardi   University of Strathclyde, UK
Andrea Bacciu   Sapienza University of Rome, Italy
Mossaab Bagdouri   Walmart Global Tech, USA

Evgenia Christoforou	CYENS Centre of Excellence, Cyprus
Abu Nowshed Chy	University of Chittagong, Bangladesh
Charles Clarke	University of Waterloo, Canada
Stephane Clinchant	Naver Labs Europe, France
Fabio Crestani	Università della Svizzera Italiana (USI), Switzerland
Shane Culpepper	The University of Queensland, Australia
Hervé Déjean	Naver Labs Europe, France
Célia da Costa Pereira	Université Côte d'Azur, France
Maarten de Rijke	University of Amsterdam, Netherlands
Arjen De Vries	Radboud University, Netherlands
Amra Deli	University of Sarajevo, Bosnia and Herzegovina
Gianluca Demartini	The University of Queensland, Australia
Danilo Dess	Leibniz Institute for the Social Sciences, Germany
Emanuele Di Buccio	University of Padua, Italy
Gaël Dias	Normandie University, France
Vlastislav Dohnal	Masaryk University, Czechia
Gregor Donabauer	University of Regensburg, Germany
Zhicheng Dou	Renmin University of China, China
Carsten Eickhoff	University of Tübingen, Germany
Michael Ekstrand	Drexel University, USA
Dima El Zein	Université Côte d'Azur, France
David Elsweiler	University of Regensburg, Germany
Ralph Ewerth	Leibniz Universität Hannover, Germany
Michael Färber	Karlsruhe Institute of Technology, Germany
Guglielmo Faggioli	University of Padova, Italy
Fabrizio Falchi	ISTI-CNR, Italy
Zhen Fan	Carnegie Mellon University, USA
Anjie Fang	Amazon.com, USA
Hossein Fani	University of Windsor, UK
Henry Field	Endicott College, USA
Yue Feng	UCL, UK
Marcos Fernández Pichel	Universidade de Santiago de Compostela, Spain
Antonio Ferrara	Polytechnic University of Bari, Italy
Komal Florio	Università di Torino - Dipartimento di Informatica, Italy
Thibault Formal	Naver Labs Europe, France
Eduard Fosch Villaronga	Leiden University, Netherlands
Maik Fröbe	Friedrich-Schiller-Universität Jena, Germany
Giacomo Frisoni	University of Bologna, Italy
Xiao Fu	University College London, UK
Norbert Fuhr	University of Duisburg-Essen, Germany

Petra Galuščáková	University of Stavanger, Norway
Debasis Ganguly	University of Glasgow, UK
Eric Gaussier	LIG-UGA, France
Xuri Ge	University of Glasgow, UK
Thomas Gerald	Université Paris Saclay CNRS SATT LISN, France
Kripabandhu Ghosh	ISSER, India
Satanu Ghosh	University of New Hampshire, USA
Daniela Godoy	ISISTAN (CONICET - UNCPBA), Argentina
Carlos-Emiliano González-Gallardo	L3i, France
Michael Granitzer	University of Passau, Germany
Nina Grgic-Hlaca	Max Planck Institute for Software Systems, Germany
Adrien Guille	Université de Lyon, France
Chun Guo	Pandora Media LLC, USA
Shashank Gupta	University of Amsterdam, Netherlands
Matthias Hagen	Friedrich-Schiller-Universität Jena, Germany
Fatima Haouari	Qatar University, Qatar
Maram Hasanain	Qatar University, Qatar
Claudia Hauff	Spotify, Netherlands
Naieme Hazrati	Free University of Bozen-Bolzano, Italy
Daniel Hienert	Leibniz Institute for the Social Sciences, Germany
Frank Hopfgartner	Universität Koblenz, Germany
Gilles Hubert	IRIT, France
Oana Inel	University of Zurich, Switzerland
Bogdan Ionescu	Politehnica University of Bucharest, Romania
Thomas Jaenich	University of Glasgow, UK
Shoaib Jameel	University of Southampton, UK
Faizan Javed	Kaiser Permanente, USA
Olivier Jeunen	ShareChat, UK
Alipio Jorge	University of Porto, Portugal
Toshihiro Kamishima	AIST, Japan
Noriko Kando	National Institute of Informatics, Japan
Sarvnaz Karimi	CSIRO, Australia
Pranav Kasela	University of Milano-Bicocca, Italy
Sumanta Kashyapi	University of New Hampshire, USA
Christin Katharina Kreutz	Cologne University of Applied Sciences, Germany
Abhishek Kaushik	Dublin City University, Ireland
Mesut Kaya	Aalborg University Copenhagen, Denmark
Diane Kelly	University of Tennessee, USA

Jae Keol Choi	Seoul National University, South Korea
Roman Kern	Graz University of Technology, Austria
Pooya Khandel	University of Amsterdam, Netherlands
Johannes Kiesel	Bauhaus-Universität, Germany
Styliani Kleanthous	CYENS CoE & Open University of Cyprus, Cyprus
Anastasiia Klimashevskaia	University of Bergen, Italy
Ivica Kostric	University of Stavanger, Norway
Dominik Kowald	Know-Center & Graz University of Technology, Austria
Hermann Kroll	Technische Universität Braunschweig, Germany
Udo Kruschwitz	University of Regensburg, Germany
Hrishikesh Kulkarni	Georgetown University, USA
Wojciech Kusa	TU Wien, Austria
Mucahid Kutlu	TOBB University of Economics and Technology, Turkey
Saar Kuzi	Amazon, USA
Jochen L. Leidner	Coburg University of Applied Sciences, Germany
Kushal Lakhotia	Outreach, USA
Carlos Lassance	Naver Labs Europe, France
Aonghus Lawlor	University College Dublin, Ireland
Dawn Lawrie	Johns Hopkins University, USA
Chia-Jung Lee	Amazon, USA
Jurek Leonhardt	TU Delft, Germany
Monica Lestari Paramita	University of Sheffield, UK
Hang Li	The University of Queensland, Australia
Ming Li	University of Amsterdam, Netherlands
Qiuchi Li	University of Padua, Italy
Wei Li	University of Roehampton, UK
Minghan Li	University of Waterloo, Canada
Shangsong Liang	MBZUAI, UAE
Nut Limsopatham	Amazon, USA
Marina Litvak	Shamoon College of Engineering, Israel
Siwei Liu	MBZUAI, UAE
Haiming Liu	University of Southampton, UK
Yiqun Liu	Tsinghua University, China
Bulou Liu	Tsinghua University, China
Andreas Lommatzsch	TU Berlin, Germany
David Losada	University of Santiago de Compostela, Spain
Jesus Lovon-Melgarejo	Université Paul Sabatier IRIT, France
Alipio M. Jorge	University of Porto, Portugal
Weizhi Ma	Tsinghua University, China

Georgios Peikos	University of Milano-Bicocca, Italy
Gustavo Penha	Spotify Research, Netherlands
Marinella Petrocchi	IIT-CNR, Italy
Aleksandr Petrov	University of Glasgow, UK
Milo Phillips-Brown	University of Edinburgh, UK
Karen Pinel-Sauvagnat	IRIT, France
Florina Piroi	Vienna University of Technology, Austria
Alessandro Piscopo	BBC, UK
Marco Polignano	Università degli Studi di Bari Aldo Moro, Italy
Claudio Pomo	Polytechnic University of Bari, Italy
Lorenzo Porcaro	Joint Research Centre European Commission, Italy
Amey Porobo Dharwadker	Meta, USA
Martin Potthast	Leipzig University, Germany
Erasmo Purificato	Otto von Guericke University Magdeburg, Germany
Xin Qian	University of Maryland, USA
Yifan Qiao	University of California, USA
Georges Quénot	Laboratoire d'Informatique de Grenoble CNRS, Germany
Alessandro Raganato	University of Milano-Bicocca, Italy
Fiana Raiber	Yahoo Research, Israel
Amifa Raj	Boise State University, USA
Thilina Rajapakse	University of Amsterdam, Netherlands
Jerome Ramos	University College London, UK
David Rau	University of Amsterdam, Netherlands
Gábor Recski	TU Wien, Austria
Navid Rekabsaz	Johannes Kepler University Linz, Austria
Zhaochun Ren	Leiden University, Netherlands
Yongli Ren	RMIT University, Australia
Weilong Ren	Shenzhen Institute of Computing Sciences, China
Chiara Renso	ISTI-CNR, Italy
Kevin Roitero	University of Udine, Italy
Tanya Roosta	Amazon, USA
Cosimo Rulli	University of Pisa, Italy
Valeria Ruscio	Sapienza University of Rome, Italy
Yuta Saito	Cornell University, USA
Tetsuya Sakai	Waseda University, Japan
Shadi Saleh	Microsoft, USA
Eric Sanjuan	Avignon Université, France
Javier Sanz-Cruzado	University of Glasgow, UK
Fabio Saracco	Centro Ricerche Enrico Fermi, Italy

Harrisen Scells	Leipzig University, Germany
Philipp Schaer	TH Köln (University of Applied Sciences), Germany
Jörg Schlötterer	University of Marburg, Germany
Ferdinand Schlatt	Friedrich-Schiller-Universität Jena, Germany
Christin Seifert	University of Marburg, Germany
Giovanni Semeraro	University of Bari, Italy
Procheta Sen	University of Liverpool, UK
Ismail Sengor Altingovde	Bilkent University, Türkiye
Vinay Setty	University of Stavanger, Norway
Mahsa Shahshahani	Accenture, Netherlands
Zhengxiang Shi	University College London, UK
Federico Siciliano	Sapienza University of Rome, Italy
Gianmaria Silvello	University of Padua, Italy
Jaspreet Singh	Amazon, USA
Sneha Singhania	Max Planck Institute for Informatics, Germany
Manel Slokom	Delft University of Technology, Netherlands
Mark Smucker	University of Waterloo, Canada
Maria Sofia Bucarelli	Sapienza University of Rome, Italy
Maria Soledad Pera	TU Delft, Germany
Nasim Sonboli	Brown University, USA
Zhihui Song	University College London, UK
Arpit Sood	Meta Inc, USA
Sajad Sotudeh	Georgetown University, USA
Laure Soulier	Sorbonne Université-ISIR, France
Marc Spaniol	Université de Caen Normandie, France
Francesca Spezzano	Boise State University, USA
Damiano Spina	RMIT University, Australia
Benno Stein	Bauhaus-Universität, Germany
Nikolaos Stylianou	Information Technologies Institute, Greece
Aixin Sun	Nanyang Technological University, Singapore
Dhanasekar Sundararaman	Duke University, UK
Reem Suwaileh	Qatar University, Qatar
Lynda Tamine	IRIT, France
Nandan Thakur	University of Waterloo, Canada
Anna Tigunova	Max Planck Institute, Germany
Nava Tintarev	University of Maastricht, Germany
Marko Tkalcic	University of Primorska, Slovenia
Gabriele Tolomei	Sapienza University of Rome, Italy
Antonela Tommasel	Aarhus University, Denmark
Helma Torkamaan	Delft University of Technology, Netherlands
Salvatore Trani	ISTI-CNR, Italy

Giovanni Trappolini	Sapienza University, Italy
Jan Trienes	University of Duisburg-Essen, Germany
Andrew Trotman	University of Otago, New Zealand
Chun-Hua Tsai	University of Omaha, USA
Radu Tudor Ionescu	University of Bucharest, Romania
Yannis Tzitzikas	University of Crete and FORTH-ICS, Greece
Venktesh V	TU Delft, Germany
Alberto Veneri	Ca' Foscari University of Venice, Italy
Manisha Verma	Amazon, USA
Federica Vezzani	University of Padua, Italy
João Vinagre	Joint Research Centre - European Commission, Italy
Vishwa Vinay	Adobe Research, India
Marco Viviani	Università degli Studi di Milano-Bicocca, Italy
Sanne Vrijenhoek	Universiteit van Amsterdam, Netherlands
Vito Walter Anelli	Politecnico di Bari, Italy
Jiexin Wang	South China University of Technology, China
Zhihong Wang	Tsinghua University, China
Xi Wang	University College London, UK
Xiao Wang	University of Glasgow, UK
Yaxiong Wu	University of Glasgow, UK
Eugene Yang	Johns Hopkins University, USA
Hao-Ren Yao	National Institutes of Health, USA
Andrew Yates	University of Amsterdam, Netherlands
Fanghua Ye	University College London, UK
Zixuan Yi	University of Glasgow, UK
Elad Yom-Tov	Microsoft, USA
Eva Zangerle	University of Innsbruck, Austria
Markus Zanker	University of Klagenfurt, Germany
Fattane Zarrinkalam	University of Guelph, Canada
Rongting Zhang	Amazon, USA
Xinyu Zhang	University of Waterloo, USA
Yang Zhang	Kyoto University, Japan
Min Zhang	Tsinghua University, China
Tianyu Zhu	Beihang University, China
Jiongli Zhu	University of California San Diego, USA
Shengyao Zhuang	The University of Queensland, Australia
Md Zia Ullah	Edinburgh Napier University, UK
Steven Zimmerman	University of Essex, UK
Lixin Zou	Wuhan University, China
Guido Zuccon	The University of Queensland, Australia

Additional Reviewers

Pablo Castells
Ophir Frieder
Claudia Hauff
Yulan He
Craig Macdonald
Graham McDonald

Iadh Ounis
Maria Soledad Pera
Fabrizio Silvestri
Nicola Tonellotto
Min Zhang

Contents – Part VI

Conference and Labs of the Evaluation Forum (CLEF)

Overview of PAN 2024: Multi-author Writing Style Analysis, Multilingual Text Detoxification, Oppositional Thinking Analysis, and Generative AI Authorship Verification
Extended Abstract

Janek Bevendorff[1], Xavier Bonet Casals[2], Berta Chulvi[3], Daryna Dementieva[4], Ashaf Elnagar[5], Dayne Freitag[6], Maik Fröbe[7], Damir Korenčić[3], Maximilian Mayerl[8], Animesh Mukherjee[9], Alexander Panchenko[10], Martin Potthast[1,11], Francisco Rangel[12], Paolo Rosso[3,13], Alisa Smirnova[14], Efstathios Stamatatos[15], Benno Stein[16], Mariona Taulé[2], Dmitry Ustalov[14], Matti Wiegmann[16(✉)], and Eva Zangerle[8]

[1] Leipzig University, Leipzig, Germany
[2] Universitat de Barcelona, Barcelona, Spain
[3] Univ. Politècnica de València, València, Spain
[4] Technical University of Munich, Munich, Germany
[5] University of Sharjah, Sharjah, United Arab Emirates
[6] SRI International, Menlo Park, USA
[7] Friedrich Schiller University Jena, Jena, Germany
[8] University of Innsbruck, Innsbruck, Austria
[9] Indian Institute of Technology Kharagpur, Kharagpur, India
[10] Skolkovo Institute of Science and Technology, Moscow, Russia
[11] ScaDS.AI, Leipzig, Germany
[12] Symanto Research, Nürnberg, Spain
[13] ValgrAI - Valencian Graduate School and Research Network of Artificial Intelligence, Valencia, Spain
[14] Toloka, Lucerne, Switzerland
[15] University of the Aegean, Mitilini, Greece
[16] Bauhaus-Universität Weimar, Weimar, Germany
pan@webis.de

Abstract. The paper gives a brief overview of the four shared tasks organized at the PAN 2024 lab on digital text forensics and stylometry to be hosted at CLEF 2024. The goal of the PAN lab is to advance the state-of-the-art in text forensics and stylometry through an objective evaluation of new and established methods on new benchmark datasets. Our four tasks are: (1) multi-author writing style analysis, which we continue from 2023 in a more difficult version, (2) multilingual text detoxification, a new task that aims to translate and re-formulate text in a non-toxic way, (3) oppositional thinking analysis, a new task that aims to discriminate critical thinking from conspiracy narratives and identify their core actors, and (4) generative AI authorship verification, which formulates the detection of AI-generated text as an authorship problem,

N. Goharian et al. (Eds.): ECIR 2024, LNCS 14613, pp. 3–10, 2024.
https://doi.org/10.1007/978-3-031-56072-9_1

one of PAN's core tasks. As with the previous editions, PAN invites software submissions as easy-to-reproduce docker containers; more than 400 pieces of software have been submitted from PAN'12 through PAN'23 combined, with all recent evaluations running on the TIRA experimentation platform [8].

1 Introduction

PAN is a workshop series and a networking initiative for stylometry and digital text forensics. PAN hosts computational shared tasks on authorship analysis, computational ethics, and the originality of writing. Since the workshop's inception in 2007, we organized 64 shared tasks[1] and assembled 55 evaluation datasets[2] plus nine datasets contributed by the community.

In 2023, our four tasks concluded with 49 submissions and 35 notebook papers. The *Multi-Author Writing Style Analysis* task was revamped for 2023 with a new dataset and structured around topical heterogeneity as an indicator for difficulty. The task attracts consistent participation of high technical quality, while the problem is still relevant and offers room for improvements, hence we continue the task with only slight modifications in 2024. The trigger detection task was newly introduced in 2023 and concluded with a variety of different solutions. While we see value in continuing to refine the task and study other promising approaches, we postpone its renewal until further ground truth can be assembled. Instead, we introduce the new *Multilingual Text Detoxification* task to better align with the interest of our community on countering toxicity and in generative tasks. The profiling cryptocurrency influencer task continued a series of author profiling tasks and concluded with high attendance and satisfying technical results. Since no significant progress is expected, we replace the task with *Oppositional Thinking Analysis* to study critical thinking and conspiracy theories in online messages. The cross-discourse type authorship verification task concluded its second iteration with mixed results and limited progress. Discriminating authorship across discourse types is difficult despite the advanced methods employed by participants. We do not expect systems to improve without further theoretical deliberation, hence we discontinue the task. Instead, we focus on the new *Generative AI Authorship Verification* task, which aims to distinguish authorship between humans and generative AI—a task of high urgency. We briefly outline the upcoming tasks in the sections that follow.

2 Multi-author Writing Style Analysis

The purpose of multi-author writing style analysis is to identify the positions of authorship changes within a document. Using authors' writing style has been shown to allow segmenting documents into parts written by different authors,

---

[1] Find PAN's past shared tasks at pan.webis.de/shared-tasks.html.

[2] Find PAN's datasets at pan.webis.de/data.html.

essentially conducting an intrinsic style analysis task and paving the way for intrinsic plagiarism detection (i.e., detecting plagiarism without the use of a reference corpus).

Multi-author writing style analysis has been part of PAN since 2016. Originally, participants had to identify and group the authors of fragments of a document [18]. In 2017, participants had to assess whether a document was written by a single or multiple authors [22] and, for multi-author documents, to find the exact positions of authorship changes. In 2018, the task was relaxed to a binary classification task aiming to assess whether a document was written by one or more authors [11]. This classification was also part of the task in 2019–2021. In 2019, the classification task was extended to determine the number of authors of multi-authored documents [27]. In 2020, participants additionally had to find changes in authorship between paragraphs [26]. In 2021, participants had to find all style changes on the paragraph level and assign all paragraphs to authors [24]. In 2022, this was extended from paragraph to sentence level [25]. In 2023, the task was relaxed to paragraph level but controlled for the simultaneous change of authorship and topic.

In the 2024 edition of the writing style analysis task, we will continue balancing "real" style changes among paragraphs and the topical similarity of paragraphs as a signal of style change. We will ask participants to solve the following intrinsic style change detection task: "For a given text, find all positions of writing style change on the paragraph level" (i.e., determine whether a style change occurred for all pairs of consecutive paragraphs). This task will be carried out on three datasets with increasing topical similarity among paragraphs and hence, increasing difficulty levels: (1) "Easy dataset": The paragraphs of a document cover various topics, allowing to infer information about authorship changes based on topic changes; (2) 'Medium dataset": The number of topics covered in a document is limited. This requires approaches to focus on style changes (rather than topic changes) to solve the detection task effectively; (3) "Hard dataset": Every paragraph in the document has the same topic.

3   Multilingual Text Detoxification

Text detoxification is a subtask of text style transfer where the style of text should be changed from toxic to neutral while preserving the content. As language modeling advances, there is growing concern about the potential unintended consequences of this technology. One such concern is the possibility of harmful or biased texts, which could perpetuate negative stereotypes or misinformation [13]. This has led to a growing interest in AI safety and the need for approaches to mitigating these risks [3]. This presents a major challenge for researchers and practitioners in language model safety, who need to develop effective detoxification techniques that can be applied to many languages.

Our first contribution to the field of text detoxification was the creation of the first parallel corpus for English together with a language-agnostic collection pipeline called ParaDetox [16]. We used this pipeline to collect a Russian parallel corpus, which was used in the first shared task on text detoxification: RUSSE-2022 [5]. The participants had to train their models based on 7k parallel toxic↔neutral pairs in Russian. The evaluation was done in two setups—automatic and manual. For both setups, three main parameters were assessed: (1) style transfer accuracy (STA), (2) content similarity (SIM), and (3) fluency (FL). Models were ranked via the geometric mean of these three parameters. Lastly, we compiled the best practices in the evaluation of text style transfer models by comparing the relationship between automatic and manual assessment [15]. The essential challenge for detoxification is that corpora with toxic↔neutral are scarce and that cross-lingual transfer of detoxification knowledge to new languages is challenging, as shown by our preliminary experiments [17].

In this first edition of the shared task on multilingual text detoxification, we want to extend the covered languages by adding Ukrainian, German, Chinese, Arabic, Amharic, and Italian. We provide participants with development sets of 1,000 parallel pairs for each of these languages. In addition, we provide the best metric for automatic evaluation for each language: (1) STA: binary toxicity classifier; (2) cosine similarity based on text embeddings; (3) either binary fluency classifier or perplexity measurement for fluency depending on the resources available for languages. The challenge for participants will be to perform cross-lingual detoxification: use a small parallel corpus in each target language, the languages metric, and the large English-Russian parallel corpus and transfer knowledge from the resource-rich to the resource-poor language. We welcome the participants to explore any multilingual large language models [4,21]. For cross-lingual knowledge transfer, approaches like back-translation [7], corpus translation [23], and adapter layers [14] training can be solid baselines.

To make a fair final evaluation on the test set, we will repeat the manual evaluation pipeline from RUSSE-2022 [5] and again utilize crowd-sourcing at Toloka.ai platform for manual evaluation. The obtained manual assessments will allow again to investigate correlations between automatic and manual metrics not only for Russian and English but for all other 6 mentioned above languages. Such corpus of human vs automatic metrics can provide base more accurate toxicity classifiers, content similarities, and fluency estimation models development.

4   Oppositional Thinking Analysis

Conspiracy theories are complex narratives that attempt to explain the ultimate causes of significant events as cover plots orchestrated by secret, powerful, and malicious groups [6]. A challenging aspect of identifying conspiracy with NLP models [9] stems from the difficulty of distinguishing critical thinking from conspiratorial thinking in automatic content moderation. This distinction is vital because labeling a message as conspiratorial when it is only oppositional could drive those who were simply asking questions into the arms of the conspiracy communities.

At PAN 2024 we aim at analyzing oppositional thinking, and more concretely, at discriminating conspiracy from critical narratives from a *stylometry* perspective. The task will address two new challenges for the NLP research community:

(1) to distinguish the conspiracy narrative from other oppositional narratives that do not express a conspiracy mentality (i.e., critical thinking); and (2) to identify in online messages the key elements of a narrative that fuels the inter-group conflict in oppositional thinking. Accordingly, we propose two sub-tasks:

Sub-task 1 is a binary classification task differentiating between (1) critical messages that question major decisions in the public health domain, but do not promote a conspiracist mentality; and (2) messages that view the pandemic or public health decisions as a result of a malevolent conspiracy by secret, influential groups.

Sub-task 2 is a token-level classification task aimed at recognizing text spans corresponding to the key elements of oppositional narratives. Since conspiracy narratives are a special kind of causal explanation, we developed a span-level annotation scheme that identifies the goals, effects, agents, and the groups-in-conflict in these narratives.

For the creation of the corpus, we first manually compiled a list of 2,273 public *Telegram* channels in *English* and *Spanish* that contain oppositional non-mainstream views on the COVID-19 pandemic.

For the second task, a new fine-grained annotation scheme was developed with the goal of identifying, at the text span level, how oppositional and conspiracy narratives use inter-group conflict. The annotation will be performed for the described 5,000 binary-labeled messages per language. We identify the following six categories of narrative elements at the span level: *Agents* (the hidden power that pulls the strings of the conspiracy. In critical messages, agents are actors that design the mainstream public health policies: Government, WHO, ...); *Objectives* (parts of the narrative that answer the question "What is intended by the agents of the CT or by the promoters of the action being criticized from a critical thinking perspective?"); *Consequences* (parts of the narrative that describe the effects of the agent's actions); *Facilitators* (the facilitators are those who collaborate with the conspirators; in critical messages, facilitators are those who implement the measures dictated by the authorities); *Campaigners* (in conspiracy messages, the campaigners are the ones who uncover the conspiracy theory; in critical messages, campaigners are those who resist the enforcement of laws and health instructions; and *Victims*, the people who are deceived into following the conspiratorial plan or the ones who suffer due to the decisions of the authorities.

5 Voight-Kampff Generative AI Authorship Verification

Authorship verification is a fundamental task in author identification. All cases of questioned authorship can be decomposed into a series of verification instances, be it in a closed-set or open-set scenario [12]. Recent editions of PAN studied authorship verification from a *cross-domain* perspective [1, 2, 20] with very high validation rates in recent years [1,2]. The latest two editions of this task studied the still challenging *cross-discourse type verification* setting [19]. Since PAN has been continuously organizing Authorship verification tasks since 2011, we are in a prime position to investigate a currently ubiquitous challenge of the highest societal importance: identifying and attributing the authorship of large language models in contrast to human authors.

Together with the ELOQUENT [10] lab on the evaluation of generative language model quality, we organize a collaborative shared task in the builder-breaker style. PAN's participants will *build* systems to detect the authorship of language models and distinguish them from human-authored texts. ELOQUENT's participants will attempt to *break* these systems by constructing evaluation datasets designed to challenge the discriminative capabilities of the PAN participants' systems.

In the builder task, participants will develop authorship verification models to attribute a text to a human or a large language model. The texts, as produced by the breakers, may, for example, source a selection of human-authored texts and modify or re-recreate them from sections, or use style transfer or error injection to make the generated text more human-like.

When including machine-generated text in authorship identification, the most desirable and hardest problem formulation is generative AI detection, where a single document is disputed without reference (see Fig. 1, Task 7). It is unclear if this detection task can be solved; it is an escalation of the standard verification setting where the authorship of two documents is decided. We hence have defined a range of problems with raising difficulty level, where the possibilities in the assignment space are in different ways constrained. In the "easiest" problem (see Fig. 1, Task 1), two documents with unknown authorship are given and we guarantee that exactly one is generated by a human, [A], and a machine, [M], respectively. This constraint is relaxed for the other tasks where, for example, both texts may also stem from a machine, { [M], [M] }. Note that an additional level of difficulty can be introduced by restricting the text lengths.

Input / Task	Possible Assignment Patterns
1. { [?], [?] }	1. { [A], [M] }
2. { [?], [?] }	2. { [A], [M] }, { [A], [A] }
3. { [?], [?] } $\longrightarrow$	3. { [A], [M] }, { [M], [M] }
4. { [?], [?] }	4. { [A], [M] }, { [A], [A] }, { [M], [M] }
5. { [?], [?] }	5. { [A], [M] }, { [A], [A] }, { [A], [B] }
6. { [?], [?] }	6. { [A], [M] }, { [A], [A] }, { [A], [B] }, { [M], [M] }
7. [?]	7. [A], [M]

Fig. 1. Hierarchy of authorship verification problems from "easy" (1) to "hard" (7), involving LLM-generated text. The difficulty results from the possible assignment patterns that are allowed to occur. [M] denotes LLM-generated text, while [A] and [B] denote human-authored text, where the same letter encodes the same human author.

Acknowledgments. The work of members of the Universitat Politècnica de València is in the framework of the XAI-DisInfodemics research project on eXplainable AI for disinformation and conspiracy detection during infodemics, funded by MCIN/AEI and by European Union NextGenerationEU/PRTR ((Grant PLEC2021-007681). The work of members of the Bauhaus-Universität Weimar and Leipzig University was partially supported by the European Commission under grant agreement GA 101070014 (https://openwebsearch.eu)

References

1. Bevendorff, J., et al.: Overview of PAN 2021: authorship verification, profiling hate speech spreaders on twitter, and style change detection. In: Experimental IR Meets Multilinguality, Multimodality, and Interaction - 12th International Conference of the CLEF Association, vol. 12880, pp. 419–431 (2021)
2. Bevendorff, J., et al.: Overview of PAN 2020: authorship verification, celebrity profiling, profiling fake news spreaders on twitter, and style change detection. In: Experimental IR Meets Multilinguality, Multimodality, and Interaction - 11th International Conference of the CLEF Association, vol. 12260, pp. 372–383 (2020)
3. Brundage, M., et al.: The malicious use of artificial intelligence: Forecasting, prevention, and mitigation. CoRR abs/1802.07228 (2018)
4. Costa-jussà, M.R., et al.: No language left behind: Scaling human-centered machine translation. arXiv e-prints pp. arXiv-2207 (2022)
5. Dementieva, D., et al.: RUSSE-2022: findings of the first Russian detoxification task based on parallel corpora. In: Computational Linguistics and Intellectual Technologies (2022)
6. Douglas, K.M., Sutton, R.M.: What are conspiracy theories? A definitional approach to their correlates, consequences, and communication. Annu. Rev. Psychol. **74**(1), 271–298 (2023). https://doi.org/10.1146/annurev-psych-032420-031329
7. El-Alami, F.Z., El Alaoui, S.O., Nahnahi, N.E.: A multilingual offensive language detection method based on transfer learning from transformer fine-tuning model. J. King Saud Univ. Comput. Inf. Sci. **34**(8), 6048–6056 (2022)
8. Fröbe, M., et al.: Continuous integration for reproducible shared tasks with TIRA.io. In: Kamps, J., et al. Advances in Information Retrieval. ECIR 2023. LNCS, vol. 13982. Springer, Cham (2023). https://doi.org/10.1007/978-3-031-28241-6_20
9. Giachanou, A., Ghanem, B., Rosso, P.: Detection of conspiracy propagators using psycho-linguistic characteristics. J. Inf. Sci. **49**(1), 3–17 (2023). https://doi.org/10.1177/0165551520985486
10. Karlgren, J., Dürlich, L., Gogoulou, E., Guillou, L., Nivre, J., Talman, A.: ELOQUENT CLEF shared tasks for evaluation of generative language model quality. In: Advances in Information Retrieval: 46th European Conference on IR Research (ECIR) (2024)
11. Kestemont, M., et al.: Overview of the author identification task at PAN 2018: cross-domain authorship attribution and style change detection. In: CLEF 2018 Labs and Workshops, Notebook Papers (2018)
12. Koppel, M., Winter, Y.: Determining if two documents are written by the same author. J. Am. Soc. Inf. Sci. **65**(1), 178–187 (2014)
13. Kumar, S., Balachandran, V., Njoo, L., Anastasopoulos, A., Tsvetkov, Y.: Language generation models can cause harm: so what can we do about it? An actionable survey. CoRR abs/2210.07700 (2022)

14. Lai, H., Toral, A., Nissim, M.: Multilingual pre-training with language and task adaptation for multilingual text style transfer. In: Muresan, S., Nakov, P., Villavicencio, A. (eds.) Proceedings of the 60th Annual Meeting of the Association for Computational Linguistics (Volume 2: Short Papers), ACL 2022, Dublin, Ireland, 22–27 May 2022, pp. 262–271, Association for Computational Linguistics (2022)
15. Logacheva, V., et al.: A study on manual and automatic evaluation for text style transfer: the case of detoxification. In: Proceedings of the 2nd Workshop on Human Evaluation of NLP Systems (HumEval), pp. 90–101, Association for Computational Linguistics, Dublin, Ireland (2022)
16. Logacheva, V., et al.: ParaDetox: detoxification with parallel data. In: Proceedings of the 60th Annual Meeting of the Association for Computational Linguistics (Volume 1: Long Papers), pp. 6804–6818, Association for Computational Linguistics, Dublin, Ireland (2022)
17. Moskovskiy, D., Dementieva, D., Panchenko, A.: Exploring cross-lingual text detoxification with large multilingual language models. In: Proceedings of the 60th Annual Meeting of the Association for Computational Linguistics: Student Research Workshop, pp. 346–354, Association for Computational Linguistics, Dublin, Ireland (2022)
18. Rosso, P., Rangel, F., Potthast, M., Stamatatos, E., Tschuggnall, M., Stein, B.: Overview of PAN2016–New Challenges for Authorship Analysis: Cross-genre Profiling, Clustering, Diarization, and Obfuscation. In: Experimental IR Meets Multilinguality, Multimodality, and Interaction. 7th International Conference of the CLEF Initiative (CLEF 16) (2016)
19. Stamatatos, E., et al.: Overview of the authorship verification task at PAN 2022. In: Faggioli, G., Ferro, N., Hanbury, A., Potthast, M. (eds.) CLEF 2022 Labs and Workshops, Notebook Papers, CEUR-WS.org (2022)
20. Stamatatos, E., Potthast, M., Pardo, F.M.R., Rosso, P., Stein, B.: Overview of the PAN/CLEF 2015 evaluation lab. In: Experimental IR Meets Multilinguality, Multimodality, and Interaction, vol. 9283, pp. 518–538 (2015)
21. Tang, Y., et al.: Multilingual translation with extensible multilingual pretraining and finetuning (2020)
22. Tschuggnall, M., et al.: Overview of the author identification task at PAN 2017: style breach detection and author clustering. In: CLEF 2017 Labs and Workshops, Notebook Papers (2017)
23. Wadud, M.A.H., Mridha, M.F., Shin, J., Nur, K., Saha, A.K.: Deep-BERT: transfer learning for classifying multilingual offensive texts on social media. Comput. Syst. Sci. Eng. **44**(2), 1775–1791 (2023)
24. Zangerle, E., Mayerl, M., Potthast, M., Stein, B.: Overview of the style change detection task at PAN 2021. In: Faggioli, G., Ferro, N., Joly, A., Maistro, M., Piroi, F. (eds.) CLEF 2021 Labs and Workshops, Notebook Papers, CEUR-WS.org (2021)
25. Zangerle, E., Mayerl, M., Potthast, M., Stein, B.: Overview of the style change detection task at PAN 2022. In: Faggioli, G., Ferro, N., Hanbury, A., Potthast, M. (eds.) CLEF 2022 Labs and Workshops, Notebook Papers, CEUR-WS.org (2022)
26. Zangerle, E., Mayerl, M., Specht, G., Potthast, M., Stein, B.: Overview of the style change detection task at PAN 2020. In: CLEF 2020 Labs and Workshops, Notebook Papers (2020)
27. Zangerle, E., Tschuggnall, M., Specht, G., Stein, B., Potthast, M.: Overview of the style change detection task at PAN 2019. In: CLEF 2019 Labs and Workshops, Notebook Papers (2019)

The CLEF 2024 Monster Track: One Lab to Rule Them All

Nicola Ferro[1], Julio Gonzalo[2(✉)], Jussi Karlgren[3], and Henning Müller[4,5]

[1] University of Padova, Padua, Italy
[2] UNED, Madrid, Spain
julio@lsi.uned.es
[3] SiloGen, Helsinki, Finland and Stockholm, Sweden
[4] HES-SO Valais, Valais, Switzerland
[5] University of Geneva, Geneva, Switzerland

Abstract. Generative *Artificial Intelligence (AI)* and *Large Language Models (LLMs)* are revolutionizing technology and society thanks to their versatility and applicability to a wide array of tasks and use cases, in multiple media and modalities. As a new and relatively untested technology, LLMs raise several challenges for research and application alike, including questions about their quality, reliability, predictability, veracity, as well as on how to develop proper evaluation methodologies to assess their various capacities.

This evaluation lab will focus on a specific aspect of LLMs, namely their versatility. The CLEF Monster Track is organized as a meta-challenge across a selection of tasks chosen from other evaluation labs running in CLEF 2024, and participants will be asked to develop or adapt a generative AI or LLM-based system that will be run on all the tasks with no or minimal task adaptation. This will allow us to systematically evaluate the performance of the same LLM-based system across a wide range of very different tasks and to provide feedback to each targeted task about the performance of a general-purpose LLM system compared to systems specifically developed for the task. Since the datasets for CLEF 2024 have not yet been released publicly, we will be able to experiment with previously unseen data, thus reducing the risk of contamination, which is one of the most serious problems faced by LLM evaluation datasets.

Keywords: Generative language models · LLM · Shared task · Quality benchmarks · CLEF

1 Motivation and Objectives

Generative *Large Language Models (LLMs)*, both proprietary models such as *Generative Pre-trained Transformer (GPT)* [16,17], and open models (i.e. which provide free access to the model weights), such as *Large Language Model Meta AI (LLaMA)* and its derivatives [4,23–25] are being successfully applied to a wide range of tasks, covering multiple media and modalities.

© The Author(s), under exclusive license to Springer Nature Switzerland AG 2024
N. Goharian et al. (Eds.): ECIR 2024, LNCS 14613, pp. 11–18, 2024.
https://doi.org/10.1007/978-3-031-56072-9_2

As a consequence, LLMs attract considerable attention from the general public, from research teams and from industry. Much effort is put into investigating the various capacities of LLMs with respect to their quality, reliability, reasoning capabilities and more. Many dataset ensembles are being adapted and used to evaluate the overall performance of LLMs, but overall there are still several challenges to address. In particular (i) the evaluation is too often compromised because test data is publicly available and models have seen the ground truth data in the pre-training phase; this problem is known as *contamination*, and is severe[1][2] for details; (ii) with the goal of testing anthropomorphic properties of models – such as common sense reasoning – and linguistic competence, datasets are drifting away from current practical application challenges.

Our goal is to systematically explore how well a given LLM performs across several real-world application challenges with respect to algorithms specifically trained for each task, avoiding contamination. We are inspired by the work of Romei et al. Hromei et al. [11], who used a single monolithic LLM to participate in all 13 EVALITA tasks[3] in 2023 – the national evaluation campaign on *Natural Language Processing (NLP)* and speech tools for Italian language – including Affect Detection, Authorship Analysis, Computational Ethics, Named Entity Recognition, Information Extraction, and Discourse Coherence. Hromei et al. [11] performed a single fine-tuning with all training data from the 13 tasks, and found that their model achieved first place in 41% of the subtasks and showcased top-three performance in 64% of them, without any task-specific prompt engineering phase. We know of no similar experiments in the *Information Retrieval (IR)* field or in other large-scale evaluation campaigns, to systematically explore cross-task performance in a shared-task setup.

Therefore, the CLEF *Monster Track*[4] will be organized as a meta-challenge across a selection of tasks chosen from the other labs running in CLEF 2024 and participants will be asked to develop a generative AI/LLM-based system that will be run against all the selected tasks with no or minimal adaptation. For each targeted task we will rely on the same dataset, experimental setting, and evaluation measures adopted for that specific task. In this way, the LLM-based systems participating in the CLEF Monster Track will be comparable directly with the specialized systems participating in each targeted task.

This allows us to systematically evaluate the performance of the same LLM-based system across a wide range of very different tasks and to provide feedback to each targeted task about the performance of a general-purpose LLM system compared to systems specifically developed for the task. Moreover, since the datasets for CLEF 2024 are in large part not public, yet, we will be able to experiment with previously unseen data, thus avoiding the risk of contamination.

---

[1] See the *LM contamination index*.
[2] https://hitz-zentroa.github.io/lm-contamination.
[3] https://www.evalita.it/.
[4] https://monsterclef.dei.unipd.it/.

2   Benchmarks for Quality Assessment of Generative Language Models

There is already a considerable and varied body of work on quality assessment of generative language models with a rich selection of benchmark resources. Many of these address capacities beyond that of generating fluent and grammatically correct language. Current evaluation procedures range over common sense reasoning [1,7,15,19,22,26], world knowledge [12,14], reading comprehension [5,6], math capabilities [8], and coding tasks [3]. Some popular aggregated benchmarks are *Massive Multitask Language Understanding (MMLU)* [10], *BIG-Bench Hard (BBH)* [21] and *Artificial General Intelligence (AGI)* Eval [27]. Other examples include Chen et al. [2], with a dataset in both Chinese and English to evaluate how well LLMs avoid hallucinating by making use of a *Retrieval-Augmented Generation (RAG)*; Gao et al. [9] with a dataset for evaluating how well LLMs generate text with citations, improving factual correctness and verifiability of the generated output; Kamalloo et al. [13] with a dataset for building end-to-end generative information-seeking models that are capable of retrieving candidate quotes and generating attributed explanations; and [13] Rashkin et al. [18] with a dataset and a two-stage annotation pipeline to evaluate attribution capacity of LLMs.

As we mentioned in the introduction, two remaining challenges are that (i) evaluation can be compromised by contamination issues (since evaluation material can be seen by the model in the pretraining process), and (ii) the overarching goal of testing anthropomorphic properties and generality of systems built on generative language models may drift away from current practical application challenges.

3   Candidate CLEF Tasks

Most CLEF labs can be used for evaluation of general-purpose technologies; but the Monster Track will primarily make use of tasks where language plays an important role and where data sets are novel — in contrast with those where public data sets are used that could have been used to train the participating LLMs. A number of candidate tasks from CLEF labs are candidates for inclusion in the Monster Track and the final selection will be made collaboratively with lab organizers.

3.1   CheckThat!

CheckThat![5] is a CLEF Lab devoted to combat misinformation. The task proposed for the Monster Track is *Check-worthiness*: given a tweet, systems must determine if it contains a claim that is worth fact checking. The organizers provide English, Arabic, and Spanish datasets to be used for instruction fine-tuning.

---

[5] https://checkthat.gitlab.io/clef2024/.

3.2 Eloquent

ELOQUENT[6] is a CLEF 2024 lab devoted to the evaluation of certain quality aspects of content generated by LLMs. It intends to use LLMs to test the capacities of themselves, and is thus a good fit for the meta-lab evaluation effort. The ELOQUENT lab proposes four evaluation tasks for LLMs:

i Topical competence: Can an LLM assess itself if it is capable to process data in some application domain of interest?
ii Veracity and Hallucination: Can an LLM be used to evaluate the output of other LLMs to detect hallucinated or factually incorrect information?
iii Robustness: Will an LLM output the same content independent of input variation which is equivalent in content but non-identical in form or style?
iv Voight-Kampff task: Can an LLM be used to detect if some piece of text is written by a human author or generated by an LLM? This task will be organised in collaboration with the PAN lab at CLEF.

3.3 Exist

EXIST[7] is a lab devoted to the detection and characterization of sexism in online content. Three tasks are proposed for the Monster Track:

i Sexism identification: given a tweet, systems must determine if it has sexist content or not.
ii Source intention: systems must determine if the sexist content is reported, judgemental or direct.
iii Sexism categorization: systems must classify sexist content into one of five categories of sexism (ideological and inequality, stereotyping and dominance, objectification, sexual violence, misogyny and non-sexual violence).

The dataset contains tweets in English and Spanish: more than 3 200 tweets per language for the training set, around 500 per language for the development set, and nearly 1 000 tweets per language for the test set. A crucial characteristic of the dataset is that it provides six annotations per tweet and task, with each annotator belonging to one out of six cohorts (three age groups × two genders). All raw annotations are provided to participants, instead of a merged single ground truth; and all annotations in the test collection are considered in the evaluation process.

3.4 ImageCLEF

ImageCLEF[8] aims to provide an evaluation forum for the cross-language annotation and retrieval of images. For the Monster Track, ImageCLEF will provide two image caption tasks in the biomedical domain (radiological images):

---

[6] https://eloquent-lab.github.io/.
[7] http://nlp.uned.es/exist2024.
[8] https://www.imageclef.org/2024.

i Concept detection where systems must predict a set of concepts (defined by the UMLS CUIs) based on the visual information provided by the radiology images.
ii Caption prediction which requires systems to automatically generate captions for the radiology images provided.

Both will use the ImageCLEFmedical 2024 Caption dataset, which consists of radiologic images of 7 different imaging modalities (angiography, CT, MRI, PET, ultrasound, X-ray, and combined modalities) with varying image dimensions as extracted from PubMed Open Access publications, along with the pre-processed image caption and a set of UMLS concepts.

3.5 LongEVAL

LongEval[9] is a shared task evaluating the temporal persistence of Information Retrieval systems and text classifiers. Two tasks are proposed:

i LongEval Retrieval: Retrieval systems are evaluated in terms of their retrieval effectiveness when the test documents are dated either right after (short term) or three months (long term) after the documents available in the train collection. The Longeval Websearch collection relies on a large set of data (corpus of pages, queries, user interaction) provided by a commercial search engine (Qwant).
ii LongEval Classification: Classification systems are evaluated in terms of their short-term effectiveness (test documents are dated shortly after training documents) and long-term effectiveness (test documents are dated more than one year apart from the training data).

3.6 PAN

The *PAN*[10] lab has organised numerous CLEF tasks related to authorship identification and verification, author profiling, plagiarism detection, and related tasks. This year PAN hosts four tasks, and two of them have been proposed to join the Monster Track effort:

i Multilingual Text Detoxification: Given a toxic piece of text in one of 7 languages, re-write it in a non-toxic way while preserving the content.
ii Voight-Kampff task (in collaboration with the ELOQUENT lab, see description above).

---

[9] https://clef-longeval.github.io/.
[10] https://pan.webis.de/.

3.7   Touché

Touché[11] is a series of scientific events and shared tasks on computational argumentation and causality. Three tasks have been proposed for the Monster Track:

i Human Value Detection: given a long text (in one of eight languages), for each sentence, identify which human values the sentence refers to and their level of attainment. The task employs a collection of roughly 3000 human-annotated texts between 400 and 800 words. The annotated values are those of the Schwartz' value continuum [20].

ii Ideology and Power Identification in Parliamentary Debates: given a parliamentary speech in one of several languages, identify the ideology of the speaker's party, and whether the speaker's party is currently governing or in opposition. The data for this task comes from ParlaMint[12], a multilingual collection of parliamentary debates.

iii Image Retrieval for Arguments. Given an argument, create a prompt for a text-to-image generator to generate an image that helps to convey the argument's premise. Organizers provide access to a Stable-Diffusion API for image generation.

3.8   Final Selection Procedure and Meta-evaluation

Selection of tasks to include in Monster Track will be made from the above list of 21 candidate tasks using criteria such as:

i Suitability: it should be possible to address every Monster Track task using a single system based on a specific LLM: we assume that participants will have a limited time to adapt their systems to each of the proposed tasks, and this is in keeping with the objective to test the generality of a system based on a generative language model.

ii Diversity: we want Monster Track tasks to cover much of the broad variety exhibited by practical challenges in information access.

iii Contamination: the test sets for Monster Track tasks should have not been made available in the past, in order to eliminate or at least minimize contamination (the possibility that language models have been exposed to the ground truth in the pre-training phase).

Details on the evaluation procedure are yet to be decided, and we will focus on qualitative insights rather than crude competition. In any case, we will rank systems with at least two procedures: (i) the average effectiveness across tasks (once all official metrics from each task are mapped into the same scale); (ii) the average rank in each of the tasks. This second procedure is more informative, as it compares the Monster Track systems with all other dedicated systems in each of the tasks. We can average the rank at least in two ways: directly (average of the

---

[11] https://touche.webis.de/.
[12] https://www.clarin.eu/parlamint.

rank or the inverse rank) or via percentiles. Percentiles have the advantage that relativise a rank in terms of the number of elements in the rank, and give more credit to, e.g., a winner with 20 opponents than to a winner with 2 opponents.

References

1. Bisk, Y., Zellers, R., Gao, J., Choi, Y., et al.: PIQA: reasoning about physical commonsense in natural language. In: Proceedings of the AAAI Conference on Artificial Intelligence, vol. 34, pp. 7432–7439 (2020)
2. Chen, J., Lin, H., Han, X., Sun, L.: Benchmarking large language models in retrieval-augmented generation (2023). arXiv.org, Computation and Language (cs.CL) arXiv:2309.01431
3. Chen, M., et al.: Evaluating large language models trained on code (2021). arXiv preprint arXiv:2107.03374
4. Chiang, W.L., et al.: Vicuna: an open-source chatbot impressing GPT-4 with 90% chatGPT quality (2023). https://lmsys.org/blog/2023-03-30-vicuna/
5. Choi, E., et al.: QuAC: question answering in context (2018). arXiv preprint arXiv:1808.07036
6. Clark, C., Lee, K., Chang, M.W., Kwiatkowski, T., Collins, M., Toutanova, K.: BoolQ: Exploring the surprising difficulty of natural yes/no questions (2019). arXiv preprint arXiv:1905.10044
7. Clark, P., et al.: Think you have solved question answering? Try ARC, the AI2 reasoning challenge (2018). arXiv preprint arXiv:1803.05457
8. Cobbe, K., et al.: Training verifiers to solve math word problems (2021). arXiv preprint arXiv:2110.14168
9. Gao, T., Yen, H., Yu, J., Chen, D.: Enabling large language models to generate text with citations (2023). arXiv.org, Computation and Language (cs.CL) arXiv:2305.14627
10. Hendrycks, D., et al.: Measuring massive multitask language understanding (2020). arXiv preprint arXiv:2009.03300
11. Hromei, C.D., Croce, D., Basile, V., Basili, R.: ExtremITA at EVALITA 2023: multi-task sustainable scaling to large language models at its extreme. In: Lai, M., Menini, S., Polignano, M., Russo, V., Sprugnoli, R., Venturi, G. (eds.) Proceedings 8th Evaluation Campaign of Natural Language Processing and Speech Tools for Italian (EVALITA 2023), CEUR Workshop Proceedings (CEUR-WS.org), ISSN 1613–0073. (2023). https://ceur-ws.org/Vol-3473/
12. Joshi, M., Choi, E., Weld, D.S., Zettlemoyer, L.: TriviaQA: a large scale distantly supervised challenge dataset for reading comprehension (2017). arXiv preprint arXiv:1705.03551
13. Kamalloo, E., Jafari, A., Zhang, X., Thakur, N., Lin, J.: HAGRID: a human-LLM collaborative dataset for generative information-seeking with attribution (2023). arXiv.org, Computation and Language (cs.CL) arXiv:2307.16883
14. Kwiatkowski, T., et al.: Natural questions: a benchmark for question answering research. Trans. Assoc. Comput. Linguist. **7**, 453–466 (2019)
15. Mihaylov, T., Clark, P., Khot, T., Sabharwal, A.: Can a suit of armor conduct electricity? A new dataset for open book question answering (2018). arXiv preprint arXiv:1809.02789
16. OpenAI: GPT-4 Technical Report (2023). arXiv.org, Computation and Language (cs.CL) arXiv:2303.08774

17. Ouyang, L., et al.: Training language models to follow instructions with human feedback. In: Koyejo, S., Mohamed, S., Agarwal, A., Belgrave, D., Cho, K., Oh, A. (eds.) Proceedings 36th Annual Conference on Neural Information Processing Systems (NeurIPS 2022) (2022). https://proceedings.neurips.cc/paper_files/paper/2022

18. Rashkin, H., et al.: Measuring attribution in natural language generation models. Comput. Linguist. 1–64 (2023)

19. Sakaguchi, K., Bras, R.L., Bhagavatula, C., Choi, Y.: WinoGrande: an adversarial WinoGrad schema challenge at scale. Commun. ACM **64**(9), 99–106 (2021)

20. Schwartz, S.H.: An overview of the Schwartz theory of basic values. Online Readings Psychol. Culture **2**(1), 11 (2012)

21. Srivastava, A., et al.: Beyond the imitation game: Quantifying and extrapolating the capabilities of language models (2022). arXiv preprint arXiv:2206.04615

22. Talmor, A., Herzig, J., Lourie, N., Berant, J.: CommonsenseQA: A question answering challenge targeting commonsense knowledge (2018). arXiv preprint arXiv:1811.00937

23. Taori, R., et al.: Alpaca: A Strong, Replicable Instruction-Following Model (2023). https://crfm.stanford.edu/2023/03/13/alpaca.html

24. Touvron, H., et al.: LLaMA: Open and Efficient Foundation Language Models (2023). arXiv.org, Computation and Language (cs.CL) arXiv:2302.13971

25. Touvron, H., et al.: Llama 2: Open Foundation and Fine-Tuned Chat Models (2023). arXiv.org, Computation and Language (cs.CL) arXiv:2307.09288

26. Zellers, R., Holtzman, A., Bisk, Y., Farhadi, A., Choi, Y.: HellaSwag: Can a machine really finish your sentence? (2019) arXiv preprint arXiv:1905.07830

27. Zhong, W., et al.: AGIEval: A human-centric benchmark for evaluating foundation models (2023). arXiv preprint arXiv:2304.06364

LifeCLEF 2024 Teaser: Challenges on Species Distribution Prediction and Identification

Alexis Joly[1]([✉])(iD), Lukáš Picek[8](iD), Stefan Kahl[6,11](iD), Hervé Goëau[2](iD),
Vincent Espitalier[2], Christophe Botella[1](iD), Benjamin Deneu[1](iD),
Diego Marcos[1](iD), Joaquim Estopinan[1], Cesar Leblanc[1], Théo Larcher[1],
Milan Šulc[10], Marek Hrúz[8](iD), Maximilien Servajean[7](iD), Jiří Matas[12],
Hervé Glotin[3](iD), Robert Planqué[4](iD), Willem-Pier Vellinga[4](iD),
Holger Klinck[6](iD), Tom Denton[9], Andrew M. Durso[13], Ivan Eggel[5],
Pierre Bonnet[2](iD), and Henning Müller[5](iD)

[1] Inria, LIRMM, Univ Montpellier, CNRS, Montpellier, France
alexis.joly@inria.fr
[2] CIRAD, UMR AMAP, Montpellier, Occitanie, France
[3] Univ. Toulon, Aix Marseille Univ., CNRS, LIS, DYNI team, Marseille, France
[4] Xeno-canto Foundation, Amersfoort, The Netherlands
[5] Informatics Institute, HES-SO Valais, Sierre, Switzerland
[6] K. Lisa Yang Center for Conservation Bioacoustics, Cornell Lab of Ornithology,
Cornell University, Ithaca, USA
[7] LIRMM, AMIS, Univ Paul Valéry Montpellier, Univ Montpellier, CNRS,
Montpellier, France
[8] Department of Cybernetics, FAV, University of West Bohemia, Pilsen, Czechia
[9] Google Research, San Francisco, USA
[10] Second Foundation, Prague, Czech Republic
[11] Chemnitz University of Technology, Chemnitz, Germany
[12] Czech Technical University, Prague, Czechia
[13] Department of Biological Sciences, Florida Gulf Coast University, Fort Myers,
USA

Abstract. Building accurate knowledge of the identity, the geographic
distribution and the evolution of species is essential for the sustainable
development of humanity, as well as for biodiversity conservation. However, species identification and inventory is a difficult and costly task,
requiring large-scale automated approaches. The LifeCLEF lab has been
promoting and evaluating advances in this domain since 2011 through
the organization of multi-year challenges. The 2024 edition presented
in this article proposes five data-driven challenges as a continuation of
this effort: (i) BirdCLEF: bird species recognition in audio soundscapes,
(ii) FungiCLEF: fungi recognition beyond 0-1 cost, (iii) GeoLifeCLEF:
remote sensing based prediction of species, (iv) PlantCLEF: Multi-
species identification in vegetation plot images, and (v) SnakeCLEF:
snake recognition in medically important scenarios.

Keywords: biodiversity · machine learning · AI · species · identification · prediction · species distribution model

1   Introduction

Accurately identifying and inventorying species is a difficult task requiring high levels of expertise and costly efforts. Since 1992, this problem has been recognized as one of the major obstacles to the global implementation of the Convention on Biological Diversity [1]. Automated approaches have thus been recognized as one of the most promising approaches since 2004 [8]. Since then, automated species identification has progressed a lot, in particular due to recent advances in deep learning [3,5,9,10,21,23,28–31]. However, even the best models remain uncertain because of the strong ambiguities between species and the scarcity of data for most species [7,22]. The LifeCLEF lab has been evaluating advances in this domain since 2014 and publishes an annual synthesis of the best methods and their performance [11–20]. Building on this effort, LifeCLEF 2024 consists of five challenges (PlantCLEF, BirdCLEF, GeoLifeCLEF, SnakeCLEF, FungiCLEF), which we briefly introduce in this paper.

2   PlantCLEF 2024 Challenge: Multi-species Plant Identification in Vegetation Plot Images

Motivation: Vegetation plot inventories are essential for ecological studies, enabling standardized sampling, biodiversity assessment, long-term monitoring and remote, large-scale surveys. They provide valuable data on ecosystems, biodiversity conservation, and evidence-based environmental decision-making. Plot images are typically one square meter in size, and botanists meticulously identify all the species found there. In addition, they quantify species abundance using indicators such as biomass, qualification factors, and areas occupied in photographs. The integration of AI could significantly improve specialists' efficiency, helping them extend the scope and coverage of ecological studies.

Data Collection: The test set will be a compilation of several image datasets of plots in different floristic contexts, such as Pyrenean and Mediterranean floras, all produced by experts. The training set will be composed more conventionally of observations of individual plants, such as those used in previous editions of PlantCLEF. More precisely, it will be a subset of the PlantCLEF2023 data focused on Europe and covering 15k plant species. It will contain about 1 million images with trusted labels (aggregated from the GBIF platform) and as many images with potentially noisy labels aggregated through web scraping (based on Google and Bing search engines).

Task Description: The main difficulty of the task lies in the shift between the test data (high-resolution multi-label images of vegetation plots) and the training data (single-label images of individual plants). The task will be evaluated as a multi-label classification task that aims to predict all the plant species on high-resolution plot images. The participants will first have access to the training set, and a few months later, they will be provided with the whole test set. Self-supervised, semi-supervised or unsupervised approaches will be strongly encouraged, and a starter package with pre-trained models will be provided. The used metric will be the mean Average Precision (computed columnwise).

3 FungiCLEF 2024 Challenge: Revisiting Fungi Recognition Beyond 1-0 Cost

Motivation: Automatic recognition of species at scale, such as in popular citizen science projects [26,27], requires efficient prediction on limited resources. In practice, species identification typically depends not solely on the visual observation of the specimen but also on other information available to the observer, e.g., habitat, substrate, location, and time. The challenge aims to provide a major benchmark for combining visual information with side information thanks to rich metadata, precise annotations, and baselines available to all competitors. Since mushrooms are often picked for consumption, the competition also considers different scenarios for misclassifying edible and poisonous mushrooms.

Data Collection: This year's challenge directly follows up on FungiCLEF 2023 [24], keeping the same training dataset, i.e. the DanihFungi2020 [25]. The data originates from a citizen science project, the Atlas of Danish Fungi, where all samples went through an expert validation process, guaranteeing high-quality labels. Rich metadata (e.g., habitat, substrate, timestamp, location, EXIF, etc.) are provided for most samples. The training set includes 295,938 training images belonging to 1,604 species observed mostly in Denmark). Validation and test datasets will cover a similar number of fungi observations (collection of images and metadata), images, and species but will reflect different periods and cover all seasons. The validation set contains 60k observations with 120k images of 3k species: 1k known from the training set and 2k unknown species. The private test set from last year will be reused. Besides, we will include a new test set originating from the newly created mobile app – CheckFungi – which includes fungi observations mainly from the Czech Republic. Such a setup will allow direct comparison with last year and measure generalization to different locations.

Task Description: The challenge aims to provide a major benchmark for combining visual observations with other observed information thanks to rich metadata, species-accurate annotations, and provided baselines. The goal of the task is to create a classification model that returns a ranked list of predicted species for a set of real fungi species observations and minimizes the danger to human life, i.e., the confusion between poisonous and edible species. The classification model will have to fit limits for memory footprint and a prediction time limit within a given Hugging Face server instance. The FungiCLEF 2024 challenge will use several metrics representing different decision scenarios, where the goal is to minimize the empirical loss L for decisions $q(x)$ over observations x and true labels y, given a cost function $W(y, q(x))$.

$$L = \sum_i W(k_i, q(x_i)) \tag{1}$$

Different recognition scenarios and their cost function $W(y, q(x))$ are described together with their motivation in last year's overview (e.g., [24]).

4 GeoLifeCLEF 2024 Challenge: Species Presence Prediction Based on Occurrences Data and High-Resolution Remote Sensing Images

Motivation: Predicting species presence in an area is vital for ecology and biodiversity conservation. These predictions inform decisions about threatened species, land use planning, protected areas, and eco-friendly agriculture. However, species distribution is influenced by complex local factors that are hard to measure, such as population interactions, landscape connectivity, habitat history, and biases in data collection. Traditional ecological models struggle to account for these factors, leading to coarse-scale resolutions. Additionally, many species are rarely observed due to sampling biases. GeoLifeCLEF aims to evaluate models on an unprecedented scale, covering thousands of species, with a spatial resolution of about 10 meters, and utilizing millions of occurrence data.

Data Collection: The 2023 edition of GeoLifeCLEF [6] revealed significant room for dataset improvement. It emphasized the need for more standardized data to enhance predictions in diverse contexts. Therefore, the 2024 edition will introduce new presence-absence data from different parts of Europe thanks to partners from the European Vegetation Archive (EVA) and the network of a large-scale European project on biodiversity monitoring (MAMBO, Horizon EU program). Test sites will be balanced better to represent the diversity of European habitats and regions, and the final test set is expected to comprise several tens of thousands of presence-absence data. Like in 2023, the training data will consist of 5 million non-standardized occurrences from GBIF spanning 38 European countries and over ten thousand plant species. Explanatory variables will include 2023's data with high-resolution remote sensing (e.g., Sentinel-2 RGB-NIR, multi-band Landsat time-series, ASTER elevation raster) and coarser resolution environmental data (e.g., Chelsa climate, SoilGrids, MODIS land use, human footprint, ISRIC soil salinity raster).

Task Description: Given a test set of geolocation and year combinations (plot) and given the corresponding high-resolution remote sensing images and environmental covariates, the goal of the task will be to return for each plot the set of species that were inventoried at that location and time by botanical experts over a small area (about 100 m^2). The test set will include only locations for which an exhaustive plant species inventory is available (i.e., in the form or presence/absence data).

5 BirdCLEF 2024 Challenge: Bird Species Identification in Soundscape Recordings

Motivation: Recognizing bird sounds in complex soundscapes is an important sampling tool that often helps reduce the limitations of point counts. In the future, archives of recorded soundscapes will become increasingly valuable as the habitats in which they were recorded will be lost. In the past few years, deep learning approaches have transformed the field of automated soundscape analysis. Yet, when training data is sparse, detection systems struggle to recognize rare species. The goal of this competition is to establish training and test datasets that can serve as real-world applicable evaluation scenarios for endangered habitats and help the scientific community to advance their conservation efforts through automated bird sound recognition.

Data Collection: We will build on the experience from previous editions and adjust the overall task to encourage participants to focus on few-shot learning and task-specific model designs. We will select training and test data to suit this demand. As in previous iterations, Xeno-canto will be the primary source for training data, and expertly annotated soundscape recordings will be used for testing. We will focus on bird species for which there is limited training data, but we will also include common species so that participants can train good recognition systems. In search of suitable test data, we will consider different data sources with varying complexity (call density, chorus, signal-to-noise ratio, anthropophony...), and quality (mono and stereo recordings). We also want to focus on specific real-world use cases (e.g., conservation efforts in India) and frame the competition based on the demand of the particular use case. Additionally, we are considering including unlabeled data to encourage self-supervised learning.

Task Description: The challenge will be held on Kaggle and the evaluation mode will resemble the 2023 test mode (i.e., hidden test data, code competition). We will use established metrics like F1 score and cmAP, which reflect use cases for which precision is key and also allow organizers to assess system performance independent of fine-tuned confidence thresholds. Participants will be asked to return a list of species for short audio segments extracted from labeled soundscape data. In the past, we used 5-second segments, and we will consider increasing the duration of these context windows to reflect the overall ground truth label distribution better. However, the overall structure of the task will remain unchanged, as it provides a well-established base that has resulted in significant participation in past editions (e.g., 1,397 participants and 21,519 submissions in 2023). Again, we will strive to keep the dataset size reasonably small (<50 GB) and easy to process, and we will also provide introductory code repositories and write-ups to lower the entry-level of the competition.

6   SnakeCLEF 2024 Challenge: Revisiting Snake Species Identification in Medically Important Scenarios

Motivation: Developing a robust system for identifying species of snakes from photographs is an important goal in biodiversity and global health. With over half a million victims of death and disability from venomous snakebite annually, understanding the global distribution of the >4000 species of snakes and differentiating species from images (particularly images of low quality) will significantly improve epidemiology data and treatment outcomes. We learned from previous editions that machines can make accurate predictions (macro averaged F1 of around 90%, and accuracy of around 90%) even in scenarios with long-tailed distributions and 1800 species [4]. However, machines still perform poorly on data from neglected regions. Thus, testing over specific countries (primarily tropical and subtropical) and integrating the medical importance of species is the next step that should provide a more reliable machine prediction.

Data Collection: The development dataset from the previous year [24] will be re-used. The dataset covers 1,784 snake species from around the world, with a minimum of three observations (i.e., multiple images of the same specimen) per species. Additionally, country-species and venomous-species mapping will be provided. The evaluation will be carried out on the same datasets as last year in order to allow direct comparison. Additionally, we will enrich the private test dataset with new data from additional neglected regions to allow testing generalization capabilities. The region won't be known to the participants. Since the competition test set is composed of private images with highly restricted licenses from individuals and natural history museums, the dataset will be undisclosed, and participants will not have access to this data.

Task Description: The SnakeCLEF challenge aims to be a major benchmark for observation-based snake species identification. The goal of the task is to create a classification model that returns a ranked list of predicted species for each set of images and location (i.e., snake observation) and minimize the danger to human life and the waste of antivenom if a bite from the snake in the image were treated as coming from the top-ranked prediction. The classification model must fit limits for memory footprint and a prediction time limit within a given Hugging Face server instance. Like last year, the 2024 edition will extend the evaluation beyond the 0-1 loss (common in classification) to motivate research in recognition scenarios with uneven costs (e.g., mistaking a venomous snake for a harmless one and vice versa). We will use two custom metrics (further described in last year's overview [24]). The first metric includes the overall classification rate (macro averaged F1) and the venomous species confusion errors. The second metric is a sum of (L) over all test observations.

7   Conclusion

The joint efforts of the LifeCLEF evaluation campaign aim to accomplish several critical goals: (i) promoting non-incremental contribution from participants, (ii) accurately measuring consistent performance gaps, (iii) gradually scaling up the problems, and (iv) enabling the growth of an engaged community. The 2024 edition aligns with this vision, offering challenges enriched with new data, new machine learning tasks (such as the one on vegetation plots), and new evaluation methodologies (such as the code-based submissions through Hugging Face imposing time and memory limits). All information about the timeline and participation in the challenges is provided on the LifeCLEF 2024 web page [2].

Acknowledgements. This work has received funding from the European Union's Horizon research and innovation program under grant agreement No. 101060639 (MAMBO project).

References

1. Convention on Biodiversity. https://www.cbd.int/
2. LifeCLEF. http://www.lifeclef.org/
3. Banan, A., Nasiri, A., Taheri-Garavand, A.: Deep learning-based appearance features extraction for automated carp species identification. Aquacult. Eng. **89**, 102053 (2020)
4. Bolon, I., Picek, L., Durso, A.M., Alcoba, G., Chappuis, F., Ruiz de Castañeda, R.: An artificial intelligence model to identify snakes from across the world: Opportunities and challenges for global health and herpetology. PLOS Neglected Tropical Diseases **16**(8), e0010647 (2022)
5. Bonnet, P., et al.: Plant identification: experts vs. machines in the era of deep learning. In: Multimedia Tools and Applications for Environmental & Biodiversity Informatics, pp. 131–149. Springer (2018)
6. Botella, C., et al.: Overview of geolifeclef 2023: species composition prediction with high spatial resolution at continental scale using remote sensing. Working Notes of CLEF (2023)
7. Garcin, C., et al.: Pl@ntnet-300k: a plant image dataset with high label ambiguity and a long-tailed distribution. In: NeurIPS 2021–35th Conference on Neural Information Processing Systems (2021)
8. Gaston, K.J., O'Neill, M.A.: Automated species identification: why not? Philosophical Trans. Roy. Soc. London B Biol. Sci. **359**(1444), 655–667 (2004)
9. Ghazi, M.M., Yanikoglu, B., Aptoula, E.: Plant identification using deep neural networks via optimization of transfer learning parameters. Neurocomputing **235**, 228–235 (2017)
10. Goodwin, A., et al.: Mosquito species identification using convolutional neural networks with a multitiered ensemble model for novel species detection. Sci. Rep. **11**(1), 13656 (2021)
11. Joly, A., et al.: Overview of lifeclef 2023: evaluation of ai models for the identification and prediction of birds, plants, snakes and fungi. In: International Conference of the Cross-Language Evaluation Forum for European Languages, pp. 416–439. Springer (2023)
12. Joly, A., et al.: Overview of LifeCLEF 2018: a large-scale evaluation of species identification and recommendation algorithms in the era of ai. In: Jones, G.J., et al. (eds.) CLEF: Cross-Language Evaluation Forum for European Languages. Experimental IR Meets Multilinguality, Multimodality, and Interaction, vol. LNCS. Springer, Avigon, France, September 2018

13. Joly, A., et al.: Overview of LifeCLEF 2019: Identification of Amazonian Plants, South & North American Birds, and Niche Prediction. In: Crestani, F., et al. (eds.) CLEF 2019 - Conference and Labs of the Evaluation Forum. Experimental IR Meets Multilinguality, Multimodality, and Interaction, vol. LNCS, pp. 387–401. Lugano, Switzerland, September 2019. https://doi.org/10.1007/978-3-030-28577-7_29. https://hal.umontpellier.fr/hal-02281455

14. Joly, A., et al.: LifeCLEF 2016: multimedia life species identification challenges. In: Fuhr, N., et al. (eds.) CLEF 2016. LNCS, vol. 9822, pp. 286–310. Springer, Cham (2016). https://doi.org/10.1007/978-3-319-44564-9_26

15. Joly, A., et al.: LifeCLEF 2017 lab overview: multimedia species identification challenges. In: Jones, G.J.F., et al. (eds.) CLEF 2017. LNCS, vol. 10456, pp. 255–274. Springer, Cham (2017). https://doi.org/10.1007/978-3-319-65813-1_24

16. Joly, A., et al.: LifeCLEF 2014: multimedia life species identification challenges. In: Kanoulas, E., et al. (eds.) CLEF 2014. LNCS, vol. 8685, pp. 229–249. Springer, Cham (2014). https://doi.org/10.1007/978-3-319-11382-1_20

17. Joly, A., et al.: LifeCLEF 2015: multimedia life species identification challenges. In: Mothe, J., Savoy, J., Kamps, J., Pinel-Sauvagnat, K., Jones, G.J.F., SanJuan, E., Cappellato, L., Ferro, N. (eds.) CLEF 2015. LNCS, vol. 9283, pp. 462–483. Springer, Cham (2015). https://doi.org/10.1007/978-3-319-24027-5_46

18. Joly, A., et al.: Overview of LifeCLEF 2020: a system-oriented evaluation of automated species identification and species distribution prediction. In: Arampatzis, A., et al. (eds.) CLEF 2020. LNCS, vol. 12260, pp. 342–363. Springer, Cham (2020). https://doi.org/10.1007/978-3-030-58219-7_23

19. Joly, A., et al.: Overview of lifeclef 2022: an evaluation of machine-learning based species identification and species distribution prediction. In: International Conference of the Cross-Language Evaluation Forum for European Languages, pp. 257–285. Springer (2022). https://doi.org/10.1007/978-3-031-13643-6_19

20. Joly, A., et al.: Overview of lifeclef 2021: an evaluation of machine-learning based species identification and species distribution prediction. In: International Conference of the Cross-Language Evaluation Forum for European Languages, pp. 371–393. Springer (2021)

21. Lee, S.H., Chan, C.S., Remagnino, P.: Multi-organ plant classification based on convolutional and recurrent neural networks. IEEE Trans. Image Process. **27**(9), 4287–4301 (2018)

22. Lorieul, T.: Uncertainty in predictions of deep learning models for fine-grained classification. Ph.D. thesis, Université Montpellier (2020)

23. Norouzzadeh, M.S., Morris, D., Beery, S., Joshi, N., Jojic, N., Clune, J.: A deep active learning system for species identification and counting in camera trap images. Methods Ecol. Evol. **12**(1), 150–161 (2021)

24. Picek, L., Chamidullin, R., Hruz, M., Durso, A.M.: Overview of fungiclef 2023: Fungi recognition beyond 1/0 cost. In: Working Notes of CLEF 2023 - Conference and Labs of the Evaluation Forum. CEUR-WS (2023)

25. Picek, L., et al.: Danish fungi 2020 - not just another image recognition dataset. In: Proceedings of the IEEE/CVF Winter Conference on Applications of Computer Vision (WACV), January 2022

26. Picek, L., Šulc, M., Matas, J., Heilmann-Clausen, J., Jeppesen, T.S., Lind, E.: Automatic fungi recognition: deep learning meets mycology. Sensors **22**(2), 633 (2022)

27. Sulc, M., Picek, L., Matas, J., Jeppesen, T., Heilmann-Clausen, J.: Fungi recognition: a practical use case. In: Proceedings of the IEEE/CVF Winter Conference on Applications of Computer Vision, pp. 2316–2324 (2020)

28. Van Horn, G., et al.: The inaturalist species classification and detection dataset. CVPR (2018)
29. Villon, S., Mouillot, D., Chaumont, M., Subsol, G., Claverie, T., Villéger, S.: A new method to control error rates in automated species identification with deep learning algorithms. Sci. Rep. **10**(1), 1–13 (2020)
30. Wäldchen, J., Mäder, P.: Machine learning for image based species identification. Methods Ecol. Evol. **9**(11), 2216–2225 (2018)
31. Wäldchen, J., Rzanny, M., Seeland, M., Mäder, P.: Automated plant species identification-trends and future directions. PLoS Comput. Biol. **14**(4), e1005993 (2018)

CLEF 2024 SimpleText Track
Improving Access to Scientific Texts for Everyone

Liana Ermakova[1(✉)] [iD], Eric SanJuan[2] [iD], Stéphane Huet[2] [iD],
Hosein Azarbonyad[3] [iD], Giorgio Maria Di Nunzio[4] [iD], Federica Vezzani[4],
Jennifer D'Souza[5] [iD], Salomon Kabongo[6] [iD], Hamed Babaei Giglou[5] [iD],
Yue Zhang[7] [iD], Sören Auer[5] [iD], and Jaap Kamps[8] [iD]

[1] Université de Bretagne Occidentale, HCTI, Brest, France
liana.ermakova@univ-brest.fr
[2] Avignon Université, LIA, Avignon, France
[3] Elsevier, Amsterdam, The Netherlands
[4] University of Padua, Padua, Italy
[5] TIB Leibniz Information Centre for Science and Technology, Hannover, Germany
[6] L3S Research Center, Leibniz University of Hannover, Hanover, Germany
[7] Technische Universität Berlin, Berlin, Germany
[8] University of Amsterdam, Amsterdam, The Netherlands
https://simpletext-project.com

Abstract. Everyone agrees on the importance of objective scientific information. However, relevant scientific documents tend to be inherently difficult to find and understand either because of intricate terminology or the potential absence of prior knowledge among their readers. Can we improve accessibility for everyone? This paper introduces the SimpleText Track at CLEF 2024, addressing the technical and evaluation challenges associated with making scientific information accessible to a wide audience, including students and non-experts. We provide appropriate reusable data and benchmarks for scientific text summarization and simplification. The CLEF 2024 SimpleText track is based on four interrelated tasks: Task 1 *Content Selection*: retrieving passages to include in a simplified summary. Task 2 *Complexity Spotting*: identifying and explaining difficult concepts. Task 3 *Text Simplification*: simplify scientific text. Task 4 *SOTA?*: tracking the state-of-the-art in scholarly publications.

Keywords: Scientific text simplification · Information extraction · Information retrieval · Natural language processing

1 Introduction

The importance of objective scientific information is universally acknowledged. In practice, accessing, processing and comprehending relevant scientific documents is challenging, due to complex terminology and the potential lack of prior knowledge among readers. The CLEF 2024 SimpleText track aims at improving accessibility to scientific information for everyone both in terms of information retrieval and natural language processing. The workshop at CLEF 2021 [3] and tracks at CLEF 2022-2023 [5,6] resulted in research community and test collections for improving access to scientific information for everyone. Specifically, test collections for retrieving relevant (and accessible) scientific text [11], for simplifying the language used in scientific documents without compromising the accuracy of the information [4], and for making complex concepts more understandable to a broader audience [2].

Scientific Text Simplification is different from traditional text simplification approaches focusing on lower literacy levels, for example making general text accessible to youth readers. Recent advances in IR and NLP hold the promise of removing some of the barriers to scientific information access.[1] The overall impact of CLEF SimpleText is to increase science literacy and broaden the audience of objective, scientific information.

The track's setup is based on the following pipeline: i) select the information to be included in a simplified summary; ii) improve the readability of the scientific text; iii) provide additional background knowledge for remaining difficult concepts; and iv) aggregate information from multiple articles. This results in the following four tasks [6]:

- *Task 1: Content Selection* retrieving passages to include in a simplified summary.
- *Task 2: Complexity Spotting* identifying and explaining difficult concepts.
- *Task 3: Text Simplification* simplify scientific text.
- *Task 4: SOTA?* tracking the state-of-the-art in scholarly publications.

In the rest of this paper, we will first reflect on the CLEF 2023 edition of the track in Sect. 2, and then provide a detailed description of each task of the CLEF 2024 edition in Sect. 3. We end with a discussion and conclusions in Sect. 4.

2   Results and Lessons from CLEF 2023 SimpleText

For the second year of running SimpleText as a track at CLEF 2023, 79 teams were registered [5]. Among them, 20 teams submitted 139 runs. For Task 1 (selecting passages/abstracts to include) [11], 39 runs were submitted by 5 teams. For Task 2 (identifying difficult terms) [2], we received 39 runs by 12 teams for subtask 2.1, and 29 runs by 10 teams for 2.2. For Task 3 (rewriting text) [4], a total of 32 submissions by 14 teams was made. The increase in active participation was encouraging.

For Task 1, we extended last year the scientific passage retrieval test collection, with a high pooling diversity, and reusable with limited pooling bias. Almost all submissions were based on neural rankers. Crossencoders and biencoders were popular approaches and turned out to be very effective. Promising results were observed for runs prioritizing credibility/complexity. This interesting feature can guide users to accessible content first, and more complex text

---

[1] A joined effort with Scholarly Document Processing https://sdproc.org/2024/.

later. In 2024, we will extend these qrels by increasing pooling depth and adding new subtopics and queries for the same set of popular science articles. We will also add supplementary labels on text complexity.

In the 2023 edition of Task 2, in addition to difficult term spotting (Task 2.1), we also asked participants to provide a definition or explanation of these concepts (Task 2.2). For the first task, both LLMs and traditional keyword extraction approaches performed well, but for the second task, LLMs outperformed traditional approaches. For Task 2.2, evaluation abbreviation or acronym expansion against ground truth is straightforward. For concepts, a set of reference sentences describing the concept was used, but many possible definitions or explanations may exist. In 2024, we will specifically evaluate the usefulness and difficulty of these explanations.

For Task 3, LLMs proved very effective in generating text simplifications, as well as for the highly complex scientific text used in the track's corpus. In general, larger models outperformed earlier models, and the limited training instances led to further improvement (but also to potential overfitting). As LLMs are used in generative mode, the analysis revealed varying degrees of hallucination where the models generated additional (and very plausible) extra content not warranted by the original input. Remarkably, this is ignored by standard evaluation measures (SARI, BLEU, ROUGE) based on text overlap with reference sentences. In 2024, we will introduce new evaluation measures that quantify unsupported content. In addition to the current sentence-based text simplification, we will also provide novel passage-based text simplification input and evaluation.

Our shared tasks are interconnected. The corpus used in Tasks 2 and 3 is based on abstracts in response to a popular science request in Task 1. In 2024, we will further expand the SimpleText test collections, provide additional evaluation measures. We will also introduce a new SOTA task aiming to generate a structured summary of scientific knowledge in multiple papers.

3   CLEF 2024 SimpleText Tasks

We will keep the three tasks from the 2023 edition and add a new one. We will reuse data constructed in previous editions with additional topics and additional automatic and manual labels. We will also emphasize automatic evaluation and training using the 2023 data.

3.1   Task 1: Retrieving Passages to Include in a Simplified Summary

Given a popular science article targeted to a general audience, this task aims at retrieving passages, which can help to understand this article, from a large corpus of academic abstracts and bibliographic metadata. Relevant passages should relate to any of the topics in the source article.

Data. We use popular science articles as a source for the types of topics the general public is interested in and as a validation of the reading level that is suitable for them. The main corpus is a large set of scientific abstracts plus associated metadata covering the field of computer science and engineering. We reuse the collection of academic abstracts from the Citation Network Dataset (12th version released in 2020)[2] [12]. This collection was extracted from DBLP, ACM, MAG (Microsoft Academic Graph), and other sources. It includes, in particular, 4,232,520 abstracts in English, published before 2020. Search requests are based on popular press articles targeted to a general audience, based on *The Guardian* and *Tech Xplore*. Each of these popular science articles represents a general topic that has to be analyzed to retrieve relevant scientific information from the corpus.

We provide the URLs to original articles, the title, and the textual content of each popular science article as a general topic. Each general topic was also enriched with one or more specific keyword queries manually extracted from their content, creating a familiar information retrieval task ranking passages or abstracts in response to a query. Available training data from 2023 includes 29 (train) and 34 (test) queries, with the later set having an extensive recall base due to the large number of submissions in 2023 [11]. In 2024, we will extend this test collection with additional test queries.

Evaluation. Topical relevance was evaluated last year with a 0–2 score on the relevance degree towards the content of the original article. In 2023, we provided an initial analysis of text complexity (based on readability measures) and authoritativeness (based on academic impact measures). In 2024, we plan to provide additional evaluation measures on both topical relevance and complexity/credibility. While these criteria can provide different levels of comparison between systems, we will continue to provide standard ranking scores based on NDCG.

3.2 Task 2: Identifying and Explaining Difficult Concepts

The goal of this task is to decide which concepts in scientific abstracts require explanation and contextualization in order to help a reader understand the scientific text. Since 2023, we ask participants to identify such concepts and to provide useful and understandable explanations for them. Thus, the task has two steps: i) to identify candidate terms in a given passage from a scientific abstract and set the level of difficulty of each term (easy or hard); ii) to provide a definition or an explanation or both only for the difficult (hard) terms.

Data. The corpus of Task 2 is based on the sentences in high-ranked abstracts to the requests of Task 1. New 2024 test data will be based on 116,763 sentences from the DBLP scientific abstracts used in Task 1.

Training data for the first step of the task, i.e. retrieving difficult terms, is based on the train and test data collected in 2023 [2]. The 2022 train data consists of 203 pairs sentence/term plus term definitions, and the 2022 test data consists of 5,142 distinct pairs sentence/term pooled from the participants' runs (1,262 distinct sentences).

---

[2] https://www.aminer.cn/citation.

Similarly, for the second step of the task, there is 2023 data available for training based on 1,000 ground truth definitions collected by Elsevier, and 5,000 mined abbreviations. The first set is extracted from a much larger corpus of full-text articles, extracted from books and articles published in ScienceDirect[3]. Moreover, there will be terminological definitions and explanations manually generated available for a subset of the training and test data used in Task 3 in 2023. A total of 175 documents and 893 sentences will be manually annotated. Finally, we encourage participants to train on existing datasets extracted from other resources such as the WCL dataset [10] to train the definition generation model, or use gazetteers, wikification resources as well as resources for abbreviation deciphering.

Evaluation. We will evaluate complex concept spotting in terms of their complexity and the detected concept spans [2]. We will automatically evaluate provided explanations by comparing them to references (e.g. ROUGE, cosine similarity, etc.). In addition, we will manually evaluate the provided explanations in terms of their usefulness with regard to a query as well as their complexity for a general audience. Note that the provided explanations can have different forms, e.g. abbreviation deciphering, examples, use cases, etc.

3.3   Task 3: Simplify Scientific Text

The goal of this task is to provide a simplified version of sentences extracted from scientific abstracts. Participants will be provided with the popular science articles and queries and matching abstracts of scientific papers, split into individual sentences.

Data. Task 3 uses the same corpus based on the sentences in high-ranked abstracts to the requests of Task 1. Our training data is a truly parallel corpus of directly simplified sentences coming from scientific abstracts from the DBLP Citation Network Dataset for *Computer Science* and Google Scholar and PubMed articles on *Health and Medicine*. Available training data from 2023 includes 648 sentences (train) and 245 sentences (test) from scientific abstracts plus manual simplifications [4]. These text passages were simplified either by master students in Technical Writing and Translation or by a domain expert (a computer scientist) and a professional translator (native English speaker) working together.

Other existing text simplification corpora used post-hoc aligned sentences [e.g., [13]. The SimpleText corpus contains 900 directly simplified sentences, and a useful addition to existing high-quality corpora like NEWSELA [13] (2,259 sentences). Our track is the first to focus on the simplification of scientific text with a much higher text complexity than news articles. In 2024, we will expand the training and evaluation data. In addition to sentence-level text simplification, we will provide passage-level input and reference simplifications, with the train and test data corresponding to 137 and 38 abstracts respectively.

---

[3] https://www.sciencedirect.com/.

Evaluation. In 2024, we will emphasize large-scale automatic evaluation measures (SARI, ROUGE, compression, readability) that provide a reusable test collection. This automatic evaluation will be supplemented with a detailed human evaluation of other aspects, essential for deeper analysis. As in 2023, we evaluate the complexity of the provided simplifications in terms of vocabulary and syntax as well as the errors (Incorrect syntax; Unresolved anaphora due to simplification; Unnecessary repetition/iteration; Spelling, typographic or punctuation errors) [4]. Almost all participants used generative models for text simplification, yet existing evaluation measures are blind to potential hallucinations with extra or distorted content [4]. In 2024, we will provide new evaluation measures that detect and quantify hallucinations in the output.

3.4 Task 4: Tracking the State-of-the-Art in Scholarly Publications

In Artificial Intelligence (AI), a common research objective is the development of new models that can report state-of-the-art (SOTA) performance. The reporting usually comprises four integral elements: Task, Dataset, Metric, and Score. These (Task, Dataset, Metric, Score) tuples coming from various AI research papers go on to power leaderboards in the community. Leaderboards, akin to scoreboards, traditionally curated by the community, are platforms displaying various AI model scores for specific tasks, datasets, and metrics. Examples of such platforms include the benchmarks feature on the Open Research Knowledge Graph and Papers with Code (PwC). Utilizing text mining techniques allows for a transition from the conventional community-based leaderboard curation to an automated text mining approach. Consequently, the goal of Task 4: SOTA? is to develop systems which given the full text of an AI paper, are capable of recognizing whether an incoming AI paper indeed reports model scores on benchmark datasets, and if so, to extract all pertinent (Task, Dataset, Metric, Score) tuples presented within the paper.

Data. The training and test datasets for this task are derived from community-curated (T, D, M, S) annotations for thousands of AI articles available on PwC (CC BY-SA). We will utilize the dataset obtained from our prior work, specifically the PwC source downloaded on May 10, 2021 [7,8], which comprised over 7,500 articles. These articles, originally sourced from arXiv under CC-BY licenses, are available in TEI XML format, each accompanied by one or more (T, D, M, S) annotations from PwC. While our previous work employed dataset splits for two-fold cross-validation experiments, for the SimpleText Task 4, we will establish new 70/30 train/test splits, providing approximately 5,000 annotated articles for participant training. A preliminary version of our training dataset can be accessed on Github https://github.com/jd-coderepos/sota.

The test set will strategically include only those articles with TDMs seen in the training set, creating a few-shot evaluation setting. Furthermore, in our subsequent research [9], we explored a zero-shot evaluation setting, wherein the dataset contained articles with at least one T, D, or M not seen in the model's training set. Thus in addition to the few-shot evaluation, we intend to introduce a second evaluation setting for Task 4, evaluating models in a zero-shot context, for which a new test dataset will be created. Finally, ongoing efforts involve expanding the primary task corpus by incorporating approximately 1,500 articles into both the train and test sets that do not report leaderboards. These articles will be annotated with the *unknown* label. Consequently, systems developed in our shared task will have comprehensive applicability to any AI article, extracting (T, D, M, S) annotations for articles that contain them and assigning *unknown* for those that do not.

Evaluation. As discussed above, in Task 4 participant systems will be evaluated in the two evaluation settings. For **Few-shot** evaluation, trained systems will have to predict (T, D, M, S) annotations on a new collection of articles' full-text. The labels in the gold dataset will include only (T, D, M, S)'s seen at least once in training. For **Zero-shot** evaluation, the task is as above with a different collection of articles, which have (T, D, M, S) with unseen T, D, or M in the training set. In both settings, the standard recall, precision, and F-score metrics will be used to report scores to the participant systems.

4   Conclusions

This paper described the setup of the CLEF 2024 SimpleText track, which contains four interconnected tasks on scientific text summarization and simplification. Within the SimpleText track, we have already released extensive corpora and manually labeled data. First, a large corpus of over 4 million scientific abstracts that can be used for popular science. Second, scientific terms from sentences coming from scientific abstracts with manually attributed difficulty scores. Third, a parallel corpus of manually simplified sentences from scientific literature. Fourth, a parallel corpus of sentences with different types of information distortion and simplification level. Please visit the SimpleText website (http://simpletext-project.com) for more details on the track.

Acknowledgments. This track would not have been possible without the great support of numerous individuals. We want to thank in particular the colleagues and the students who participated in data construction, evaluation and reviewing. We also thank the MaDICS (https://www.madics.fr/ateliers/simpletext/) research group and the French National Research Agency (project *ANR-22-CE23-0019-01*). SimpleText's SOTA Task is jointly funded by the Deutsche Forschungsgemeinschaft (DFG, German Research Foundation) - project number: NFDI4DataScience (460234259) and the German BMBF project SCINEXT (01lS22070).

References

1. Aliannejadi, M., Faggioli, G., Ferro, N., Vlachos, M. (eds.): Working Notes of CLEF 2023: Conference and Labs of the Evaluation Forum, CEUR Workshop Proceedings, vol. 3497, CEUR-WS.org (2023). http://ceur-ws.org/Vol-3497
2. Ermakova, L., Azarbonyad, H., Bertin, S., Augereau, O.: Overview of the CLEF 2023 SimpleText Task 2: difficult concept identification and explanation. In: [1]. https://ceur-ws.org/Vol-3497/paper-239.pdf
3. Ermakova, L., et al.: Text Simplification for Scientific Information Access: CLEF 2021 SimpleText Workshop. In: Advances in Information Retrieval - 43nd European Conference on IR Research, ECIR 2021, Lucca, Italy, March 28 - April 1, 2021, Proc., Lucca, Italy (2021)
4. Ermakova, L., Bertin, S., McCombie, H., Kamps, J.: Overview of the CLEF 2023 SimpleText Task 3: Scientific text simplification. In: [1]. https://ceur-ws.org/Vol-3497/paper-240.pdf
5. Ermakova, L., SanJuan, E., Huet, S., Azarbonyad, H., Augereau, O., Kamps, J.: Overview of the CLEF 2023 SimpleText Lab: automatic simplification of scientific texts. In: Arampatzis, A., et al. (eds.) CLEF'23: Proceedings of the Fourteenth International Conference of the CLEF Association. LNCS. Springer (2023). https://doi.org/10.1007/978-3-031-42448-9_30
6. Ermakova, L., et al.: Overview of the CLEF 2022 SimpleText lab: automatic simplification of scientific texts. In: Barrón-Cedeño, A., et al. (eds.) CLEF'22: Proceedings of the Thirteenth International Conference of the CLEF Association. LNCS. Springer (2022)
7. Kabongo, S., D'Souza, J., Auer, S.: Automated mining of leaderboards for empirical ai research. In: Towards Open and Trustworthy Digital Societies: 23rd International Conference on Asia-Pacific Digital Libraries, ICADL 2021, Virtual Event, December 1–3, 2021, Proceedings 23, pp. 453–470. Springer (2021)
8. Kabongo, S., D'Souza, J., Auer, S.: Orkg-leaderboards: a systematic workflow for mining leaderboards as a knowledge graph. arXiv preprint arXiv:2305.11068 (2023)
9. Kabongo, S., D'Souza, J., Auer, S.: Zero-shot entailment of leaderboards for empirical ai research. In: Proceedings of the ACM/IEEE Joint Conference on Digital Libraries in 2023 (2023)
10. Navigli, R., Velardi, P.: Learning word-class lattices for definition and hypernym extraction. In: ACL, pp. 1318–1327 (2010)
11. SanJuan, E., Huet, S., Kamps, J., Ermakova, L.: Overview of the CLEF 2023 simpletext task 1: passage selection for a simplified summary. In: [1]. https://ceur-ws.org/Vol-3497/paper-238.pdf
12. Tang, J., Zhang, J., Yao, L., Li, J., Zhang, L., Su, Z.: ArnetMiner: extraction and mining of academic social networks. In: KDD'08, pp. 990–998 (2008)
13. Xu, W., Callison-Burch, C., Napoles, C.: Problems in current text simplification research: new data can help. Trans. ACL **3**, 283–297 (2015). ISSN 2307–387X. https://www.mitpressjournals.org/doi/abs/10.1162/tacl_a_00139

CLEF 2024 JOKER Lab: Automatic Humour Analysis

Liana Ermakova[1], Anne-Gwenn Bosser[2], Tristan Miller[3,4](✉),
Tremaine Thomas[1], Victor Manuel Palma Preciado[1,5], Grigori Sidorov[5],
and Adam Jatowt[6]

[1] Université de Bretagne Occidentale, HCTI, Brest, France
[2] École Nationale d'Ingénieurs de Brest, Lab-STICC CNRS UMR 6285, Brest, France
[3] Department of Computer Science, University of Manitoba, Winnipeg, Canada
`tristan.miller@umanitoba.ca`
[4] Austrian Research Institute for Artificial Intelligence (OFAI), Vienna, Austria
[5] Instituto Politécnico Nacional (IPN), Centro de Investigación en
Computación (CIC), Mexico City, Mexico
[6] University of Innsbruck, Innsbruck, Austria

Abstract. The JOKER Lab at the Conference and Labs of the Evaluation Forum (CLEF) aims to foster research on automated processing of verbal humour, including tasks such as retrieval, classification, interpretation, generation, and translation. Despite the heady success of large language models, humour and wordplay automatic processing are far from being a solved problem. JOKER brings together experts from the social and computational sciences and encourages them to collaborate on shared tasks with quality-controlled annotated datasets. In 2024, we will offer entirely new shared tasks on humour-aware information retrieval, as well as fine-grained sentiment analysis and classification of humour for conversational agents. As in the past JOKER Labs, we will also make our data available for an unshared task that solicits novel use cases. In this paper, we provide a brief retrospective on the JOKER Labs, with a focus on the results and lessons learnt from last year's iteration, and we preview the tasks to be held at JOKER 2024.

Keywords: Wordplay · Puns · Humour · Humour-aware information retrieval · Humour generation · Humour classification · Sentiment analysis · Information retrieval

1 Introduction

The automated analysis and processing of humour remains a challenge for artificial intelligence (AI) and natural language processing (NLP) systems, not least because of its reliance on nuanced contextual clues, lexical and syntactic ambiguities, implicit cultural references, and disregard for linguistic patterns or rules. Now in its third year, the JOKER Lab[1] at the Conference and Labs of the Evaluation Forum (CLEF) aims to foster work on this challenging and multi-faceted

---

[1] https://www.joker-project.com/.

N. Goharian et al. (Eds.): ECIR 2024, LNCS 14613, pp. 36–43, 2024.
https://doi.org/10.1007/978-3-031-56072-9_5

topic. We do this by setting out well-defined tasks, creating reusable quality-controlled data for training and evaluation, and encouraging cooperation among researchers from AI and the social sciences, particularly linguistics.

In the first edition of JOKER, held at CLEF 2022, we introduced pilot shared tasks that focussed on the categorisation, interpretation, and translation of wordplay, puns and humorous neologisms, in English and French. We also provided our data for an unshared task [11, 12].[2] In the 2023 edition, we focussed on the detection, location, and interpretation of puns in English, French, and Spanish [6, 7], as well as on machine translation of wordplay from English into French and English into Spanish [8–10]. For JOKER 2024, we keep the pun translation task and we broaden the scope of the tasks by tackling humour in general instead of wordplay in particular. We also extend the corpus we had previously constructed for pun detection [5, 9] to enable a new task on humour-aware information retrieval. Finally, we plan to offer entirely new tasks on fine-grained sentiment analysis of humour as well as its automatic classification designed to advance the analysis and generation of humour in conversational agents. Despite recent significant advances in LLMs, the production deployment of generated humour in conversational agents is still risky as such AI models might reveal undesirable stereotyping biases or produce jokes that can be perceived as offensive or inappropriate, humour often being related to social taboos [14]. On the other hand, the classification of humour is helpful in providing an appropriate answer to a playful request in dialogue systems [19].

The four shared tasks of JOKER 2024 can be summarised as follows:

Task 1 Humour-aware information retrieval
Task 2 Fine-grained sentiment analysis of short humorous texts
Task 3 Humour classification according to genre and technique
Task 4 Translation of puns from English to French

In addition to these shared tasks, we will hold our usual unshared task to attract runs on new use cases, such as the generation or evaluation of humour.

2 JOKER 2023: Results and Lessons Learnt

In the 2023 edition of JOKER, our primary focus was on wordplay in three languages: English, French, and Spanish. Specifically, we concentrated on the tasks of wordplay detection (identifying whether a pun is used in a given span of text), wordplay location (pinpointing the specific location within the text of the ambiguous words exploited for the wordplay), and wordplay interpretation (describing the double meaning of a pun). We also held machine translation tasks for English

---

[2] In a shared task, the organisers establish the criteria for assessing an unresolved artificial intelligence issue and create a dataset that has been annotated by humans for training and testing. Participants in the challenge then use the publicly available training data to create solutions for addressing the problem. Subsequently, the organisers assess these solutions using a private, undisclosed test dataset. In an unshared task, the organisers offer annotated data without specifying a specific problem to solve. Participants are encouraged to leverage this data to propose and address new, unique problems of their choosing.

to French and to Spanish. We constructed large corpora for these tasks and made them available to the scientific community [5,6,9]. We manually evaluated machine translations of English puns into French and Spanish. These machine translations provide new positive examples of wordplay in these languages as well as a large number of negative examples. Note that both positive and negative examples are similar in length and vocabulary, making them effective for AI model testing. This data can be used to further enrich our large corpora for wordplay detection in French and Spanish. We will use these examples to augment data for Task 1 (humour-aware information retrieval) in 2024.

Fifty teams registered for JOKER 2023. Of these, thirteen participated in the tasks, submitting a total of 186 runs for the tasks on pun detection (40 for English, 17 for French, and 18 for Spanish), location (English 25, French 11, Spanish 11), interpretation (English 15, French 2), and translation (20 English to French, 27 English to Spanish). Our analysis of the submitted runs, most of which were produced with large language models (LLMs), suggest that the tasks remain a challenge for state-of-the-art systems [9].

In the two previous editions of JOKER, participants in the unshared tasks used our data for a variety of new use cases, including text transformation and conversational systems, sentiment analysis, and comparison of machine and human performance for pun detection, interpretation, and translation [9,12].

3   Shared Tasks

3.1   Task 1: Humour-Aware Information Retrieval

As we have shown previously, the state-of-the-art AI models are wordplay- and humour-agnostic [5,6,9]. To foster research in humour-aware information retrieval, we introduce a new task that aims at retrieving short humorous texts from a document collection. The intended use case is to search for a joke on a specific topic. This can be useful for humour researchers in the humanities, for second-language learners as a learning aid, for professional comedians as a writing aid, and for translators who might need to adapt certain jokes to other cultures.

To construct queries we will use the locations of wordplay from JOKER 2023's Task 2 [7]. The document collection will be an extension of the collection used for JOKER 2023's tasks on wordplay detection in English and French [6] and on machine translations of puns from English into French [8]. We will augment the data by introducing retrieved text passages from humour and non-humour sources as well as generated data on topics relevant to the queries. This methodology will ensure the reduction of data artefacts related to unbalanced topics, differences in vocabulary of humour and non-humour texts, and the differences between machine- and human-produced documents. The application of data augmentation techniques should also prevent participants' use of the previously released data to simply mine answers or perform web scraping, reducing in this way the reliance on external sources and overfitting existing corpora. We note that the detection of machine-generated texts is also of limited utility.

We will use standard information retrieval metrics to evaluate participants' systems for this task, such as MAP, NDCG, precision, and recall.

3.2 Task 2: Fine-Grained Sentiment Analysis of Short Humorous Texts

Despite significant advances of foundational and generative LLMs, humour generation is still a challenge [20]. Although some commercial organisations have carried out research on humour detection and generation [17–19,21], the production deployment of generated humour in conversational agents is risky as such agents might reveal undesirable stereotyping biases that propagate negative stereotypes and prejudices involving gender, race, religion, and other social constructs [3,14]. Moreover, humour is often related to social taboos; according to the relief theory of humour, the comic effect is achieved by easing the strain arising from the suppression of socially unacceptable wants and cravings [15]. The superiority theory of humour suggests that humour arises when we perceive ourselves as superior to someone or something else [16]. In either case, the produced joke can be perceived as offensive or inappropriate. To overcome this issue, it is crucial to understand the potential sentiment of the interlocutor.

Our Task 2 involves a nuanced examination of the sentiment of short humorous texts, aiming to identify and classify various emotional nuances, including positive, negative, and neutral sentiments. The challenge is to go beyond traditional two- or three-way polarity classification and instead distinguish among subtle variations in sentiment, thus providing a deeper understanding of the emotional intricacies of humour. The motivation for this task is to understand nuanced humour emotions that can enhance content recommendations, benefit creators by fine-tuning humour, aid cross-cultural communication, improve content moderation, and enhance educational tools for emotional intelligence development.

Besides the usual polarity labels (positive/negative/neutral), we will ask participating systems to predict classical emotions such as anger, surprise, disgust, enjoyment, fear, and sadness [4] that can occur as a reaction to the joke. Such emotions are often related to the meaning or pragmatic goals of the humour – for example:

– Humour can help defuse *anger* by offering a different perspective, highlighting the absurdity of a situation, or providing a comedic outlet for expressing frustration.
– Humour thrives on *surprising* the audience and subverting expectations.
– Humour can transform *disgust* by using exaggeration, absurdity, or satire to highlight the repulsive elements in a comedic context.
– Humour is inherently tied to *enjoyment*.
– Humour can help alleviate *fear* by providing relief and a sense of safety.
– Humour can bring levity to difficult situations, offering a fresh perspective or a momentary escape from *sadness*.

We will also ask systems to label jokes that propagate negative generalisations involving social constructs.

We will collect short humorous texts, including jokes, puns, and one-liners sourced from a diverse range of publicly available humour corpora as well as from the JOKER pun detection subcorpus in English. The dataset is designed to encompass various forms of humour, linguistic styles, and thematic content.

Systems will be evaluated using traditional metrics for multi-label classification in IR (precision, recall, F_1, and accuracy).

3.3 Task 3: Humour Classification

Classification of humour is an important task in dialogue systems as it can be used to provide an appropriate answer to a playful request [19]. Following recent work in this area [18], we will classify humorous texts according to the main theories of humour: the *relief theory*, the *incongruity theory*, and the *superiority theory*. The relief theory has been used to explain adult and scatological humour – for example, shopping requests in the form of incongruously humorous orders that are physically or legally impossible, out of scope, characterised by unreasonable quantity, or regarding improbable topics[3] – and superiority theory may explain requests related to the personification of chatbots like Alexa or that include cultural references. As we plan to use a dataset which is not limited to virtual assistant requests, we will adapt this classification to further types of humour. We will consider puns as a special case of incongruity. We will also include irony and satire in the dataset.

Texts will be subject to automatic classification according to the following humour techniques:

Irony relies on a gap between the literal meaning and the intended meaning, creating a humorous twist or reversal.

Sarcasm involves using irony to mock, criticize, or convey contempt.

Exaggeration involves magnifying or overstating something beyond its normal or realistic proportions.

Incongruity refers to the unexpected or contradictory elements that are combined in a humorous way.

Absurdity involves presenting situations, events, or ideas that are inherently illogical, irrational, or nonsensical.

Self-deprecating humour involves making fun of oneself or highlighting one's own flaws, weaknesses, or embarrassing situations in a lighthearted manner.

Wit refers to clever, quick, and intelligent humour.

Surprise in humour involves introducing unexpected elements, twists, or punchlines that catch the audience off guard.

Thus, the humour classification of Task 3 is a classification where the goal is to identify in a target text the particular technique used for generating humour. As

---

[3] The classification proposed in the related past work [18] was intended to classify playful requests to the Amazon Alexa virtual assistant.

in Task 2, runs for this task will be evaluated according to standard metrics for classification.

The data for this task will be a mixture of existing corpora on irony and sarcasm detection [1,13] and on COVID-19 humour [2], as well as jokes retrieved from public humour sites according to the predefined categories selected in a balanced manner. An example data instance is given below:

Sentence "Finally figured out the reason I look so bad in photos. It's my face."
Humour technique Self-deprecating

3.4   Task 4: Pun Translation

We continue the pun translation task as in JOKER 2022 [12] and 2023 [9]. The goal of this task is to translate English punning jokes into French. Translations should aim to preserve, to the extent possible, both the form and meaning of the original wordplay. We will provide an updated training and test set of English-French translations of punning jokes and continue the practice of having trained experts manually evaluate system translations according to features such as lexical field preservation, sense preservation, wordplay form preservation, style shift, humorousness shift, etc. and the presence of errors in syntax, word choice, etc. Runs will be ranked according to the number of successful translations – i.e., translations preserving, to the extent possible, both the form and sense of the original wordplay. We will also experiment with other semi-automatic metrics.

4   Conclusion

This paper has presented a brief retrospective overview of previous editions of the JOKER Lab, highlighting the key aspects of past tasks. We have furthermore outlined the prospective setup of the CLEF 2024 JOKER Lab, which features shared tasks on humour-aware information retrieval and two tasks potentially useful for humour generation in dialogue systems, namely fine-grained sentiment analysis of short humorous texts and humour classification. We also welcome submissions using our data for other research tasks, such as for pun generation, for humour perception, or as input for virtual animated characters. Prospective participants are encouraged to visit the JOKER website at https://joker-project. com for further details on the lab.

Acknowledgments. This project has received a government grant managed by the National Research Agency under the program "Investissements d'avenir" integrated into France 2030, with the Reference ANR-19-GURE-0001. JOKER is supported by *La Maison des sciences de l'homme en Bretagne*. This Lab would not have been possible without the great support of numerous individuals; we would like to thank in particular the students of the Université de Bretagne Occidentale for their contribution to data construction.

References

1. Abu Farha, I., Oprea, S.V., Wilson, S., Magdy, W.: SemEval-2022 Task 6: iSarcasm-Eval, intended sarcasm detection in English and Arabic. In: Proceedings of the 16th International Workshop on Semantic Evaluation (SemEval-2022), pp. 802–814. Association for Computational Linguistics, July 2022. https://doi.org/10.18653/v1/2022.semeval-1.111

2. Bogireddy, N.R., Suresh, S., Rai, S.: I'm out of breath from laughing! I think? A dataset of COVID-19 humor and its toxic variants. In: Companion Proceedings of the ACM Web Conference 2023, pp. 1004–1013. Association for Computing Machinery, New York, NY (2023). https://doi.org/10.1145/3543873.3587591

3. Chulvi, B., Rosso, P., Labadie Tamayo, R.: Everybody hurts, sometimes: overview of HUrtful HUmour at IberLEF 2023: Detection of humour spreading prejudice in Twitter. Procesamiento del lenguaje natural (71), 383–395 (2023). https://dialnet.unirioja.es/servlet/articulo?codigo=9093263

4. Ekman, P.: Emotions Revealed: Recognizing Faces and Feelings to Improve Communication and Emotional Life. Henry Holt and Company, March 2004

5. Ermakova, L., Bosser, A.G., Jatowt, A., Miller, T.: The JOKER Corpus: English-French parallel data for multilingual wordplay recognition. In: SIGIR '23: Proceedings of the 46th International ACM SIGIR Conference on Research and Development in Information Retrieval, pp. 2796–2806. Association for Computing Machinery, New York, NY (2023). https://doi.org/10.1145/3539618.3591885

6. Ermakova, L., Miller, T., Bosser, A.G., Palma Preciado, V.M., Sidorov, G., Jatowt, A.: Overview of JOKER 2023 Automatic Wordplay Analysis Task 1 - pun detection. In: Aliannejadi, M., Faggioli, G., Ferro, N., Vlachos, M. (eds.) Working Notes of CLEF 2023 - Conference and Labs of the Evaluation Forum. CEUR Workshop Proceedings, vol. 3497, pp. 1785–1803 (Oct 2023)

7. Ermakova, L., Miller, T., Bosser, A.G., Palma Preciado, V.M., Sidorov, G., Jatowt, A.: Overview of JOKER 2023 Automatic Wordplay Analysis Task 2 - pun location and interpretation. In: Aliannejadi, M., Faggioli, G., Ferro, N., Vlachos, M. (eds.) Working Notes of CLEF 2023 - Conference and Labs of the Evaluation Forum. CEUR Workshop Proceedings, vol. 3497, pp. 1804–1817 (Oct 2023)

8. Ermakova, L., Miller, T., Bosser, A.G., Palma Preciado, V.M., Sidorov, G., Jatowt, A.: Overview of JOKER 2023 Automatic Wordplay Analysis Task 3 - pun translation. In: Aliannejadi, M., Faggioli, G., Ferro, N., Vlachos, M. (eds.) Working Notes of CLEF 2023 - Conference and Labs of the Evaluation Forum. CEUR Workshop Proceedings, vol. 3497, pp. 1818–1827 (Oct 2023)

9. Ermakova, L., Miller, T., Bosser, A.G., Palma Preciado, V.M., Sidorov, G., Jatowt, A.: Overview of JOKER - CLEF-2023 track on automatic wordplay analysis. In: Arampatzis, A., et al. (eds.) Experimental IR Meets Multilinguality, Multimodality, and Interaction, vol. 14163, pp. 397–415. Springer Nature Switzerland, Cham (2023). https://doi.org/10.1007/978-3-031-42448-9_26

10. Ermakova, L., Miller, T., Bosser, A.G., Palma Preciado, V.M., Sidorov, G., Jatowt, A.: Science for fun: the CLEF 2023 JOKER track on automatic wordplay analysis. In: Kamps, J., et al. (eds.) Advances in Information Retrieval: 45th European Conference on Information Retrieval, ECIR 2023, Dublin, Ireland, April 2–6, Proceedings, Part III. Lecture Notes in Computer Science, vol. 13982, pp. 546–556. Springer, Heidelberg (Apr 2023). https://doi.org/10.1007/978-3-031-28241-6_63

11. Ermakova, L., Miller, T., Puchalski, O., Regattin, F., Mathurin, É., Araújo, S., Bosser, A.-G., Borg, C., Bokiniec, M., Corre, G.L., Jeanjean, B., Hannachi, R., Mallia, G, Matas, G., Saki, M.: CLEF Workshop JOKER: Automatic Wordplay and Humour Translation. In: Hagen, M., Verberne, S., Macdonald, C., Seifert, C., Balog, K., Nørvåg, K., Setty, V. (eds.) ECIR 2022. LNCS, vol. 13186, pp. 355–363. Springer, Cham (2022). https://doi.org/10.1007/978-3-030-99739-7_45

12. Ermakova, L., et al.: Overview of JOKER@CLEF 2022: automatic wordplay and humour translation workshop. In: Barrón-Cedeño, A., et al. (eds.) Experimental IR Meets Multilinguality, Multimodality, and Interaction. Proceedings of the Thirteenth International Conference of the CLEF Association (CLEF 2022). LNCS, vol. 13390, pp. 447–469 (2022). https://doi.org/10.1007/978-3-031-13643-6_27

13. Frenda, S., Pedrani, A., Basile, V., Lo, S.M., Cignarella, A.T., Panizzon, R., Marco, C., Scarlini, B., Patti, V., Bosco, C., Bernardi, D.: EPIC: multi-perspective annotation of a corpus of irony. In: Proceedings of the 61st Annual Meeting of the Association for Computational Linguistics, vol. 1, pp. 13844–13857. Association for Computational Linguistics, July 2023. https://doi.org/10.18653/v1/2023.acl-long.774

14. Liang, P.P., Wu, C., Morency, L.P., Salakhutdinov, R.: Towards understanding and mitigating social biases in language models. In: Meila, M., Zhang, T. (eds.) Proceedings of the 38th International Conference on Machine Learning. Proceedings of Machine Learning Research, vol. 139, pp. 6565–6576. PMLR, July 2021. https://proceedings.mlr.press/v139/liang21a.html

15. Morreall, J.: Humor, philosophy and education. Educ. Philos. Theory **46**(2), 120–131 (2014). https://doi.org/10.1080/00131857.2012.721735

16. Morreall, J.: Philosophy of humor. In: Zalta, E.N., Nodelman, U. (eds.) The Stanford Encyclopedia of Philosophy. Metaphysics Research Lab, Stanford University, summer 2023 edn. (2023). https://plato.stanford.edu/archives/sum2023/entries/humor/

17. Shani, C., Libov, A., Tolmach, S., Lewin-Eytan, L., Maarek, Y., Shahaf, D.: "Alexa, what do you do for fun?" Characterizing playful requests with virtual assistants, May 2021. http://arxiv.org/abs/2105.05571. arXiv:2105.05571 [cs]

18. Shani, C., Libov, A., Tolmach, S., Lewin-Eytan, L., Maarek, Y., Shahaf, D.: "Alexa, do you want to build a snowman?" Characterizing playful requests to conversational agents. In: CHI Conference on Human Factors in Computing Systems Extended Abstracts, pp. 1–7. ACM, New Orleans, LA, April 2022. https://doi.org/10.1145/3491101.3519870

19. Shapira, N., Kalinsky, O., Libov, A., Shani, C., Tolmach, S.: Evaluating humorous response generation to playful shopping requests. In: Kamps, J., Goeuriot, L., Crestani, F., Maistro, M., Joho, H., Davis, B., Gurrin, C., Kruschwitz, U., Caputo, A. (eds.) Advances in Information Retrieval, vol. 13981, pp. 617–626. Springer, Cham (2023). https://doi.org/10.1007/978-3-031-28238-6_53

20. Winters, T.: Computers learning humor is no joke. Harvard Data Science Review **3**(2), April 2021. https://doi.org/10.1162/99608f92.f13a2337

21. Ziser, Y., Kravi, E., Carmel, D.: Humor detection in product question answering systems. In: Proceedings of the 43rd International ACM SIGIR Conference on Research and Development in Information Retrieval, pp. 519–528. ACM, July 2020. https://doi.org/10.1145/3397271.3401077

Advancing Multimedia Retrieval in Medical, Social Media and Content Recommendation Applications with ImageCLEF 2024

Bogdan Ionescu[1,3], Henning Müller[2,3], Ana Maria Drăgulinescu[1,3],
Ahmad Idrissi-Yaghir[3,4], Ahmedkhan Radzhabov[3,16],
Alba Garcia Seco de Herrera[3,5], Alexandra Andrei[1,3(✉)], Alexandru Stan[3,6],
Andrea M. Storås[3,7], Asma Ben Abacha[3,8], Benjamin Lecouteux[3,19],
Benno Stein[3,17], Cécile Macaire[3,19], Christoph M. Friedrich[3,4],
Cynthia Sabrina Schmidt[3,10], Didier Schwab[3,19],
Emmanuelle Esperança-Rodier[3,19], George Ioannidis[3,6], Griffin Adams[3,9],
Henning Schäfer[3,10], Hugo Manguinhas[3,11], Ioan Coman[1,3],
Johanna Schöler[3,13], Johannes Kiesel[3,17], Johannes Rückert[3,4], Louise Bloch[3,4],
Martin Potthast[3,18], Maximilian Heinrich[3,17], Meliha Yetisgen[3,14],
Michael A. Riegler[3,7], Neal Snider[3,15], Pål Halvorsen[3,7], Raphael Brüngel[3,4],
Steven A. Hicks[3,7], Vajira Thambawita[3,7], Vassili Kovalev[3,12,16],
Yuri Prokopchuk[3,16], and Wen-Wai Yim[3,8]

[1] National University of Science and Technology Politehnica Bucharest, Bucharest, Romania
alexandra.andrei@upb.ro
[2] University of Applied Sciences Western Switzerland (HES-SO), Sierre, Switzerland
[3] CEA LIST, Paris, France
[4] University of Applied Sciences and Arts Dortmund, Dortmund, Germany
[5] University of Essex, Colchester, UK
[6] IN2 Digital Innovations, Lindau, Germany
[7] SimulaMet, Oslo, Norway
[8] Microsoft, Redmond, USA
[9] Columbia University, New York, USA
[10] University Hospital Essen, Essen, Germany
[11] Europeana Foundation, Hague, Netherlands
[12] Belarus State University, The Hague, Belarus
[13] Sahlgrenska University Hospital, Gothenburg, Sweden
[14] University of Washington, Seattle, USA
[15] Microsoft/Nuance, Burlington, USA
[16] Belarus National Academy of Sciences, Minsk, Belarus
[17] Bauhaus-Universität Weimar, Weimar, Germany
[18] Leipzig University and ScaDS.AI, Leipzig, Germany
[19] Univ. Grenoble Alpes, CNRS, Grenoble INP*, Grenoble, France

Abstract. The ImageCLEF evaluation campaign was integrated with CLEF (Conference and Labs of the Evaluation Forum) for more than

Apart from the general organisers, authors are listed in alphabetical order.

N. Goharian et al. (Eds.): ECIR 2024, LNCS 14613, pp. 44–52, 2024.
https://doi.org/10.1007/978-3-031-56072-9_6

20 years and represents a Multimedia Retrieval challenge aimed at evaluating the technologies for annotation, indexing, and retrieval of multimodal data. Thus, it provides information access to large data collections in usage scenarios and domains such as medicine, argumentation and content recommendation. ImageCLEF 2024 has four main tasks: (i) a *Medical* task targeting automatic image captioning for radiology images, synthetic medical images created with Generative Adversarial Networks (GANs), Visual Question Answering and medical image generation based on text input, and multimodal dermatology response generation; (ii) a joint ImageCLEF-Touché task *Image Retrieval/Generation for Arguments* to convey the premise of an argument, (iii) a *Recommending* task addressing cultural heritage content-recommendation, and (iv) a joint ImageCLEF-ToPicto task aiming to provide a translation in pictograms from natural language. In 2023, participation increased by 67% with respect to 2022 which reveals its impact on the community.

Keywords: Medical AI · image captioning · GANs · Visual Question Answering · response generation · cultural heritage · argumentation

1   Introduction

With a tradition of more than 20 years, the ImageCLEF benchmarking campaign provides the scientific community with research activities and evaluation of approaches for annotation, indexing, classification and retrieval of multimodal data. ImageCLEF 2024 is integrated with the Conference and Labs of the Evaluation Forum (CLEF) [18,19], with the 22nd edition being hosted by University of Grenoble Alpes, France, 9–12 September 2024[1]. Considering the experience of the last four successful editions, ImageCLEF 2024 will handle a diversity of applications within the four benchmarking tasks approaching different aspects of mono- and cross-language information retrieval systems [14,18,19] related to the interpretation of the radiology images [25], and testing the hypothesis that the artificial biomedical images contain fingerprints of the original images [6,12], to name a few. The campaign targets multimodal data annotation and retrieval community and researchers from computer vision, image information retrieval and digital image processing fields. From its inception, ImageCLEF demonstrated a meaningful scholarly impact and, currently, there are 420 publications with 3792 citations on Web of Science (WoS). The paper introduces the four tasks planned for 2024, namely: ImageCLEFmedical, ImageCLEFrecommeding, Image Retrieval/Generation for Arguments, and ImageCLEFtoPicto (Fig. 1).

2   ImageCLEFmedical

ImageCLEFmedical task is currently at its 20th edition [19]. The 2024 edition will continue all the medical sub-tasks in 2023, namely: (i) *caption* task with medical

---

[1] https://clef2024.imag.fr/.

concept detection and caption prediction, (ii) *GAN* task on synthetic medical images generated with GANs, (iii) *MEDVQA-GI* task for medical images generation based on text input, (iv) *Mediqa* task with a new use-case on multimodal dermatology response generation.

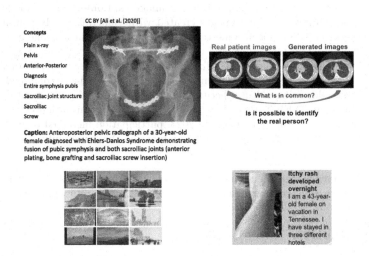

Fig. 1. Sample images from (left to right, top to bottom): ImageCLEFmedical-caption with an image with the corresponding CUIs and caption, ImageCLEFmedical-GAN with examples of real and generated images, ImageCLEFrecommending with an example of editorial "European landscapes and landmarks" Gallery, and ImageCLEF-Mediqa with example of medical details related to a skin problem and the associated image.

ImageCLEFmedical-Caption[2]. The *caption* task consists in the interpretation of the insights gained from radiology images and in this 8th edition [8,10,21–23,27,28], there are also two subtasks: concept detection and caption prediction. The *concept detection* subtask aims to develop competent systems that are able to predict the Unified Medical Language System (UMLS®) Concept Unique Identifiers (CUIs) based on the visual image content. The F1-Score [11] will be used for evaluation. The *caption prediction* subtask focuses on implementing models to predict captions for given radiology images. To improve the evaluation quality, BERTScore will be refined by integrating domain-specific models such as BioBERT [15]. The scope of scoring may be expanded experimenting with other models, such as ClinicalBLEURT and MedBERTScore [3]. In 2024, an updated version of the Radiology Objects in Context (ROCO) [24] dataset will be used, further extended with new PubMed images. As a novelty in the 8th edition, an optional, experimental explainability extension will be offered for both tasks, and participants will be asked to provide explanations (e.g., heatmaps, Shapley values) for a small subset of images manually evaluated.

---

[2] https://www.imageclef.org/2024/medical/caption.

ImageCLEFmedical-GAN[3]. The *GANs* task is a relatively new challenge in the ImageCLEFmedical track. We will continue with the first task proposed in the previous edition focused on examining the existing hypothesis that GANs generate medical images containing certain "fingerprints" of the authentic images used for generative network training. The task supposes performing analysis of test image datasets and assessing the probability with which certain images of real patients were used for training image generators. The participants will test the hypothesis on two different levels, including identifying the source dataset used for training and exploring the problem of detection and isolation of image regions in generated images that inherit the patterns presented in the original ones. The second task explores the hypothesis that generative models imprint unique fingerprints on generated images and whether different generative models or architectures leave discernible signatures within the synthetic images they produce. Similar to the previous year, the 2D gray-scale images being provided depict the axial slices of CT scans of tuberculosis patients taken at different stages of their treatment. In 2024, we continue to use the advanced Diffuse Models [17] along with other generative models for image generation.

ImageCLEFmedical-MEDVQA-GI[4]. In the 2nd MEDVQA-GI challenge, the participants have to generate medical images based on text input, along with optimal prompts for off-the-shelf generative models to improve the diagnosis and classification of real medical images using AI-generated images. The dataset is built up on the one collected in the first edition of MEDVQA-GI [13]. The task is divided into two subtasks: i) Image Synthesis (IS) requiring participants to leverage text-to-image generative models to create a rich dataset of medical images derived from textual prompts; 2) Optimal Prompt Generation (OPG) asking to generate optimal textual prompts to guide an off-the-shelf generative model in creating realistic medical images with imaging modalities ranging from magnetic resonance imaging (MRI) and computerised tomography (CT) scans to endoscopic images of various medical conditions. The subjective evaluation is made by medical experts and on how accurately a model trained on these AI-generated images can classify real medical images. The metrics employed are the Fréchet Inception Distance (FID) and standard metrics such as accuracy, precision, recall, and F1 score in both single-center and multi-center datasets.

ImageCLEFmedical-Mediqa[5]. The *MEDIQA-magic* task is a new task focusing on multimodal dermatology response generation. Participants are given a clinical narrative context along with accompanying images. They have to generate a textual response answering the health question based on the described history and images. Examples of responses include diagnosis of the problem and suggestions for treatment. The dataset is created by using real consumer health users' queries and images; gold standard responses are composed by certified medical doctors. Response output will be evaluated using metrics such as ROUGE, BLEURT, and BERTScore.

---

[3] https://www.imageclef.org/2024/medical/gans.
[4] https://www.imageclef.org/2024/medical/vqa.
[5] https://www.imageclef.org/2024/medical/mediqa.

Table 1. Submissions for the argument. The first (retrieved) image could help to convey the "forget"-part of the premise but does not relate to voting. The second image (generated) is rated higher on relevance (1 vs. 2). The third image (generated) does not indicate that someone forgets their ID or is barred from voting (irrelevant, 0).

Topic: Photo identification at polling stations
Claim: Legislation to impose restrictive photo ID requirements has the potential
           to block millions of American voters.
Premise: People will forget their ID cards and be denied their right to vote.

Images:
Rationale Relevance: 1   2   0

3 Image Retrieval/Generation for Arguments

Touché-Argument-Image[6] gives a set of arguments asking to return several images for each argument, helping to convey the argument's premise. Participants can optionally add a short rationale explaining the meaning of the image and can employ both image retrieval and image generation approaches. We provide a conclusion and a topic as context for the premise (see Table 1 for an example). There are three kinds of submissions: (1) Retrieval: Participants can retrieve suitable images from a focused crawl, where we also provide automatically recognised text from the image and text from web pages containing the image. (2) Prompted Generation: Following the idea of the infinite index [7], participants can submit prompts for the Stable Diffusion image generator [26] (we provide a stable API). (3) Direct: Participants can employ other reproducible methods to generate images (e.g., chart generators) to submit them directly. The task follows the classic TREC-style methodology with ranked results judged by human assessors. The performance will be evaluated with nDCG.

4 ImageCLEFrecommending

ImageCLEFrecommending[7] is a task which focuses on content-recommendation for cultural heritage content. Despite current advances in content-based recommendation systems, there is limited understanding of how well these perform and how relevant they are for the final end-users. This task aims to fill this gap by benchmarking different recommendation systems and methods. The task builds upon a key infrastructure for researchers and heritage professionals, namely Europeana [9] and requires participants to devise recommendation methods and systems, apply them in the supplied dataset gathered from Europeana

---

[6] https://touche.webis.de/clef24/touche24-web/image-retrieval-for-arguments.html.
[7] https://www.imageclef.org/2024/recommending.

and provide a series of recommendations for items and editorials within two sub-tasks: (i) given a list of items, provide a list of recommended items; (ii) given an editorial (Europeana blog or gallery), provide a list of recommended editorials. A new dataset based on Europeana items and editorials will be provided to the participants including metadata based on the Europeana Data Model schema. Performance will be evaluated based on the recommendations that are provided computing Mean Average Precision at X (Map@X) compared to the ground truth. Moreover, because black-box systems make it difficult for users to assess why the recommendation should be trusted, the systems providing explanation for the results will be awarded additional point.

5   ImageCLEFtoPicto

ImageCLEFtoPicto introduces two new tasks whose objective is to provide a translation in pictograms from a natural language, either from (i) text or (ii) speech understandable by the users, in this case, people with language impairments. ImageCLEFtoPicto is an opportunity for the research community who works in the field of Augmentative and Alternative Communication (AAC), and pictogram translation to gather around two challenging tasks. The task is divided into two subtasks: (i) text-to-pictogram translation focuses on automatically generating a corresponding sequence of pictogram terms from a French text, and (ii) speech-to-pictogram focuses on the two modalities of speech and pictograms. The data of the tasks are taken from Traitement de Corpus Oraux en Français (TCOF) [1], a French speech corpus. The challenge is to directly translate speech to pictogram terms without going through the transcription dimension, which is the focus of the speech community with current spoken language translation systems [4,5]. Both automatic (BLEU [20], METEOR [2], WER [29]) and human evaluation (MQM framework [16]) will be carried out by experts in the domain.

6   Conclusions

The paper highlights the tasks proposed by the 22nd edition of ImageCLEF evaluation campaign. The tasks are refreshed and the participants will meet new challenging use-cases as image retrieval/generation for argumentation or generation of optimal textual prompts to guide an off-the-shelf generative model to create realistic medical images. Moreover, the two joint tasks with Touché and ToPicto increase to a larger extent the diversity of the datasets and metrics which are shared with the community, whereas tasks such as Fusion and Aware will be discontinued. Overall, ImageCLEF2024 will continue to provide the researchers with the possibility to assess the performance of their conceived systems having access to a common evaluation framework to compare their results.

Acknowledgement. The lab is supported under the H2020 AI4Media "A European Excellence Centre for Media, Society and Democracy" project, contract #951911. The work of Louise Bloch and Raphael Brüngel was partially funded by a PhD grant from the University of Applied Sciences and Arts Dortmund (FH Dortmund), Germany. The work of Ahmad Idrissi-Yaghir and Henning Schäfer was funded by a PhD grant from the DFG Research Training Group 2535 Knowledge- and data-based personalisation of medicine at the point of care (WisPerMed). Image Retrieval/Generation for Arguments task was partially supported by the European Commission under grant agreement GA 101070014.

References

1. André, V., Canut, E.: Mise à disposition de corpus oraux interactifs: le projet tcof (traitement de corpus oraux en français). Pratiques. Linguistique, littérature, didactique (147–148), 35–51 (2010)
2. Banerjee, S., Lavie, A.: Meteor: An automatic metric for mt evaluation with improved correlation with human judgments. In: Proceedings of the ACL Workshop on Intrinsic and Extrinsic Evaluation Measures for Machine Translation and/or Summarization, pp. 65–72 (2005)
3. Ben Abacha, A., Yim, W.w., Michalopoulos, G., Lin, T.: An investigation of evaluation methods in automatic medical note generation. In: Findings of the Association for Computational Linguistics: ACL 2023, pp. 2575–2588. Association for Computational Linguistics, Toronto, Canada, July 2023. https://doi.org/10.18653/v1/2023.findings-acl.161. https://aclanthology.org/2023.findings-acl.161
4. Bérard, A., Besacier, L., Kocabiyikoglu, A.C., Pietquin, O.: End-to-end automatic speech translation of audiobooks. In: 2018 IEEE International Conference on Acoustics, Speech and Signal Processing (ICASSP), pp. 6224–6228. IEEE (2018)
5. Bérard, A., Pietquin, O., Besacier, L., Servan, C.: Listen and translate: a proof of concept for end-to-end speech-to-text translation. In: NIPS Workshop on end-to-end learning for speech and audio processing. Barcelona, Spain, December 2016. https://hal.science/hal-01408086
6. Carlini, N., et al.: Extracting training data from diffusion models (2023)
7. Deckers, N., et al.: The infinite index: information retrieval on generative text-to-image models. In: Gwizdka, J., Rieh, S.Y. (eds.) ACM SIGIR Conference on Human Information Interaction and Retrieval (CHIIR 2023), pp. 172–186. ACM, March 2023. https://doi.org/10.1145/3576840.3578327. https://doi.org/10.1145/3576840.3578327
8. Eickhoff, C., Schwall, I., García Seco de Herrera, A., Müller, H.: Overview of Image-CLEFcaption 2017 - the image caption prediction and concept extraction tasks to understand biomedical images. In: Working Notes of Conference and Labs of the Evaluation Forum (CLEF 2017). CEUR Workshop Proceedings, vol. 1866. CEUR-WS.org (2017)
9. Europeana Foundation: Europeana (2022). https://www.europeana.eu/
10. García Seco De Herrera, A., Eickhof, C., Andrearczyk, V., Müller, H.: Overview of the ImageCLEF 2018 caption prediction tasks. In: Working Notes of Conference and Labs of the Evaluation Forum (CLEF 2018). CEUR Workshop Proceedings, vol. 2125. CEUR-WS.org (2018)
11. Goutte, C., Gaussier, E.: A probabilistic interpretation of precision, recall and F-score, with implication for evaluation. In: Advances in Information Retrieval - 27th European Conference on IR Research (ECIR 2005), pp. 345–359. Springer (2005)
12. Gui, J., Sun, Z., Wen, Y., Tao, D., Ye, J.: A review on generative adversarial networks: algorithms, theory, and applications. IEEE Trans. Knowl. Data Eng. (2021). https://doi.org/10.1109/TKDE.2021.3130191

13. Hicks, S.A., Storås, A., Halvorsen, P., de Lange, T., Riegler, M.A., Thambawita, V.: Overview of ImageCLEFmedical 2023 - Medical Visual Question Answering for Gastrointestinal Tract. In: CLEF2023 Working Notes. CEUR Workshop Proceedings, CEUR-WS.org, Thessaloniki, Greece, 18–21 September 2023

14. Kalpathy-Cramer, J., García Seco de Herrera, A., Demner-Fushman, D., Antani, S., Bedrick, S., Müller, H.: Evaluating performance of biomedical image retrieval systems: overview of the medical image retrieval task at ImageCLEF 2004–2014. Computerized Med. Imaging Graph. **39**, 55–61 (2015)

15. Lee, J., Yoon, W., Kim, S., Kim, D., Kim, S., So, C.H., Kang, J.: BioBERT: a pre-trained biomedical language representation model for biomedical text mining. Bioinform. **36**, 1234–1240 (2020). https://doi.org/10.1093/bioinformatics/btz682

16. Lommel, A., Uszkoreit, H., Burchardt, A.: Multidimensional quality metrics (mqm): a framework for declaring and describing translation quality metrics. Tradumàtica **12**, 0455–0463 (2014)

17. Mueller-Franzes, G., et al.: A multimodal comparison of latent denoising diffusion probabilistic models and generative adversarial networks for medical image synthesis. Sci. Rep. **13**(1), July 2023. https://doi.org/10.1038/s41598-023-39278-0. https://doi.org/10.1038%2Fs41598-023-39278-0

18. Müller, H., Clough, P., Deselaers, T., Caputo, B. (eds.): ImageCLEF - Experimental Evaluation in Visual Information Retrieval, The Springer International Series On Information Retrieval, vol. 32. Springer, Heidelberg (2010)

19. Müller, H., Kalpathy-Cramer, J., García Seco de Herrera, A.: Experiences from the ImageCLEF medical retrieval and annotation tasks. In: Information Retrieval Evaluation in a Changing World. TIRS, vol. 41, pp. 231–250. Springer, Cham (2019). https://doi.org/10.1007/978-3-030-22948-1_10

20. Papineni, K., Roukos, S., Ward, T., Zhu, W.J.: BLEU: a method for automatic evaluation of machine translation. In: Proceedings of the 40th Annual Meeting of the Association for Computational Linguistics (ACL 2002), pp. 311–318 (2002)

21. Pelka, O., Abacha, A.B., García Seco de Herrera, A., Jacutprakart, J., Friedrich, C.M., Müller, H.: Overview of the ImageCLEFmed 2021 concept & caption prediction task. In: Working Notes of Conference and Labs of the Evaluation Forum (CLEF 2021). CEUR Workshop Proceedings, vol. 2936. CEUR-WS.org (2021)

22. Pelka, O., Friedrich, C.M., García Seco de Herrera, A., Müller, H.: Overview of the ImageCLEFmed 2019 concept detection task. In: Working Notes of Conference and Labs of the Evaluation Forum (CLEF 2019). CEUR Workshop Proceedings, vol. 2380. CEUR-WS.org (2019)

23. Pelka, O., Friedrich, C.M., García Seco de Herrera, A., Müller, H.: Overview of the ImageCLEFmed 2020 concept prediction task: Medical image understanding. In: Working Notes of Conference and Labs of the Evaluation Forum (CLEF 2020). CEUR Workshop Proceedings, vol. 2696. CEUR-WS.org (2020)

24. Pelka, O., Koitka, S., Rückert, J., Nensa, F., Friedrich, C.M.: Radiology objects in COntext (ROCO): a multimodal image dataset. In: Stoyanov, D., Taylor, Z., Balocco, S., Sznitman, R., Martel, A., Maier-Hein, L., Duong, L., Zahnd, G., Demirci, S., Albarqouni, S., Lee, S.-L., Moriconi, S., Cheplygina, V., Mateus, D., Trucco, E., Granger, E., Jannin, P. (eds.) LABELS/CVII/STENT -2018. LNCS, vol. 11043, pp. 180–189. Springer, Cham (2018). https://doi.org/10.1007/978-3-030-01364-6_20

25. Pelka, O., Nensa, F., Friedrich, C.M.: Annotation of enhanced radiographs for medical image retrieval with deep convolutional neural networks. PLoS ONE **13**(11), e0206229 (2018). https://doi.org/10.1371/journal.pone.0206229

26. Rombach, R., Blattmann, A., Lorenz, D., Esser, P., Ommer, B.: High-resolution image synthesis with latent diffusion models. CoRR abs/2112.10752 (2021). https://arxiv.org/abs/2112.10752

27. Rückert, J., et al.: Overview of ImageCLEFmedical 2022 - caption prediction and concept detection. In: CLEF2022 Working Notes. CEUR Workshop Proceedings, CEUR-WS.org, Bologna, Italy, 5–8 Sept 2022

28. Rückert, J., et al.: Overview of ImageCLEFmedical 2023 - caption prediction and concept detection. In: CLEF2023 Working Notes. CEUR Workshop Proceedings, CEUR-WS.org, Thessaloniki, Greece, 18–21 Sept 2023

29. Woodard, J., Nelson, J.: An information theoretic measure of speech recognition performance. In: Workshop on Standardisation for Speech I/O Technology, Naval Air Development Center, Warminster, PA (1982)

iDPP@CLEF 2024: The Intelligent Disease Progression Prediction Challenge

Helena Aidos[1], Roberto Bergamaschi[2], Paola Cavalla[3], Adriano Chiò[4],
Arianna Dagliati[2], Barbara Di Camillo[5], Mamede Alves de Carvalho[1],
Nicola Ferro[5(✉)], Piero Fariselli[4], Jose Manuel García Dominguez[6],
Sara C. Madeira[1], and Eleonora Tavazzi[7]

[1] University of Lisbon, Lisbon, Portugal
{haidos,sacmadeira}@fc.ul.pt, mamedemg@mail.telepac.pt
[2] University of Pavia, Pavia, Italy
roberto.bergamaschi@mondino.it, arianna.dagliati@unipv.it
[3] "Città della Salute e della Scienza", Turin, Italy
paola.cavalla@unito.it
[4] University of Turin, Turin, Italy
{adriano.chio,piero.fariselli}@unito.it
[5] University of Padua, Padua, Italy
{barbara.dicamillo,nicola.ferro}@unipd.it
[6] Gregorio Marañon Hospital in Madrid, Madrid, Spain
jgarciadominguez@salud.madrid.org
[7] IRCCS Foundation C. Mondino in Pavia, Pavia, Italy

Abstract. *Amyotrophic Lateral Sclerosis (ALS)* and *Multiple Sclerosis (MS)* are chronic diseases characterized by progressive or alternate impairment of neurological functions (motor, sensory, visual, cognitive). Patients have to manage alternated periods in hospital with care at home, experiencing a constant uncertainty regarding the timing of the disease acute phases and facing a considerable psychological and economic burden that also involves their caregivers. Clinicians, on the other hand, need tools able to support them in all the phases of the patient treatment, suggest personalized therapeutic decisions, indicate urgently needed interventions.

iDPP@CLEF run in CLEF 2022 and 2023, offering tasks on the prediction of ALS and MS progression, using retrospective patient clinical data complemented with environmental data.

iDPP@CLEF 2024 will focus on prospective patient data for ALS collected via a dedicated app developed by the BRAINTEASER project and sensor data in the context of clinical trials in Turin, Pavia, Lisbon, and Madrid. For MS, iDPP@CLEF 2024 will rely on retrospective patient data complemented with environmental and pollution data from clinical institutions in Pavia and Turin.

N. Goharian et al. (Eds.): ECIR 2024, LNCS 14613, pp. 53–59, 2024.
https://doi.org/10.1007/978-3-031-56072-9_7

1   Introduction

Amyotrophic Lateral Sclerosis (ALS) and *Multiple Sclerosis (MS)* are severe chronic diseases characterized by a progressive but variable impairment of neurological functions, characterized by high heterogeneity both in presentation features and rate of disease progression. As a consequence patients' needs are different, challenging both caregivers and clinicians. Indeed, the time of relevant events is variable, which is associated with uncertainty regarding the opportunity of critical interventions, like non-invasive ventilation and gastrostomy in the case of ALS, with implications on the quality of life of patients and their caregivers. For this reason, clinicians need tools able to support their decision in all phases of disease progression and underscore personalized therapeutic decisions. Indeed, this heterogeneity is partly responsible for the lack of effective prognostic tools in medical practice, as well as for the current absence of a therapy able to effectively slow down or reverse the disease course. On the one hand, patients need support for facing the psychological and economic burdens deriving from the uncertainty of how the disease will progress; moreover, clinicians require tools that may assist them throughout the patient's care, recommending tailored therapeutic decisions and providing alerts for urgently needed actions. We need to design and develop *Artificial Intelligence (AI)* algorithms to:

- stratify patients according to their phenotype all over the disease evolution;
- predict the progression of the disease in a probabilistic, time dependent way;
- better describe disease mechanisms.

The *Intelligent Disease Progression Prediction at CLEF (iDPP@CLEF)* lab[1] is organized by the BRAINTEASER project and aims to deliver an evaluation infrastructure for driving the development of such AI algorithms. Indeed, in this context, it is fundamental, even if not so common yet, to develop shared approaches, promote the use of common benchmarks, foster the comparability and replicability of the experiments. Differently from previous challenges in the field, iDPP@CLEF addresses in a systematic way some issues related to the application of AI in clinical practice in ALS and MS. In addition to defining risk scores based on the probability of occurrence of an event in the short or long term period, iDPP@CLEF also addresses the issue of providing information in a more structured and understandable way to clinicians.

The paper is organized as follows: Sects. 2 and 3 present what has been done in iDPP@CLEF 2022 and 2023; Sect. 4 introduces the datasets made available by iDPP@CLEF; while Sect. 5 discusses the plans for iDPP@CLEF 2024; finally, Sect. 6 draws some conclusions.

2   iDPP@CLEF 2022

iDPP@CLEF [6,7] run as a pilot lab for the first time in CLEF 2022 and focused on pilot activities aimed both at an initial exploration of ALS progression pre-

---

[1] https://brainteaser.health/open-evaluation-challenges/.

diction and at understanding the challenges and limitations to refine and tune the lab itself for future iterations.

iDPP@CLEF 2022 consisted of the following tasks:

- **Pilot Task 1 - Ranking Risk of Impairment**: it focused on ranking of patients based on the risk of impairment in specific domains. More in detail, we used the *ALS Functional Rating Scale Revisited (ALSFRS-R)* scale [1] to monitor speech, swallowing, handwriting, dressing/hygiene, walking and respiratory ability in time and asked participants to rank patients based on time to event risk of experiencing impairment in each specific domain.
- **Pilot Task 2 - Predicting Time of Impairment**: it refined Task 1 by asking participants to predict when specific impairments will occur (i.e. in the correct time-window). In this regard, we assessed model calibration in terms of the ability of the proposed algorithms to estimate a probability of an event close to the true probability within a specified time-window.
- **Position Paper Task 3 - Explainability of AI algorithms**: we evaluated proposals of different frameworks able to explain the multivariate nature of the data and the model predictions.

43 participants registered for iDPP@CLEF 2022 and 5 participants successfully submitted a total of 120 runs for Task 1 and Task 2; moreover, 2 position papers were submitted for the explainability task.

Submission of participants are openly available in git repositories[2] and all the participant papers and slides are available through the iDPP@CLEF 2022 web site[3].

3 iDPP@CLEF 2023

iDPP@CLEF 2023 [3,4] was the second iteration of the lab, expanding its scope to include both ALS and MS in the study of disease progression. The activities in iDPP@CLEF 2023 focus on two objectives: exploring the prediction of MS worsening and conducting a more in-depth analysis of ALS compared to iDPP@CLEF 2022, with the addition of environmental data.

iDPP@CLEF 2023[4] will organize the following activities:

- **Task 1 – Predicting Risk of Disease Worsening (MS)**: it focused on ranking subjects based on the risk of worsening, setting the problem as a survival analysis task. More specifically the risk of worsening predicted by the algorithm should reflect how early a patient experiences the "worsening" event, and should range between 0 and 1. Worsening is defined on the basis of the *Expanded Disability Status Scale (EDSS)* [8], accordingly to clinical standards. In particular, we considered two different definitions of worsening corresponding to two different sub-tasks:

---

[2] https://bitbucket.org/brainteaser-health/.

[3] https://brainteaser.health/open-evaluation-challenges/idpp-2022/.

[4] https://brainteaser.health/open-evaluation-challenges/idpp-2023/.

- *Subtask 1a*: the patient crosses the threshold EDSS ≥ 3 at least twice within one year interval;
- *Subtask 1b*: the second definition of worsening depends on the first recorded value accordingly to current clinical protocols. If Baseline EDSS < 1, worsening event occurs when and increase of EDSS by 1.5 points is first observed; if $1 \leq$ Baseline EDSS < 5.5, worsening event occurs when and increase of EDSS by 1 point is first observed; if baseline EDSS ≥ 5.5, worsening event occurs when and increase of EDSS by 0.5 points is first observed.

For each sub-task, participants were given a dataset containing 2.5 years of visits, with the occurrence of the worsening event and the time of occurrence pre-computed by the challenge organizers.

- **Task 2 – Predicting Cumulative Probability of Worsening (MS)**: it refined Task 1 by asking participants to explicitly assign the cumulative probability of worsening at different time windows, i.e., between years 0 and 2, 0 and 4, 0 and 6, 0 and 8, 0 and 10.

In particular, we considered two different definitions of worsening corresponding to two different sub-tasks – *subtask 2a* and *subtask 2b* – defined according to the same rules as in Task 1.

- **Position Paper Task 3 – Impact of Exposition to Pollutants (ALS)**: we evaluated proposals of different approaches to assess if exposure to different pollutants is a useful variable to predict time to *Percutaneous Endoscopic Gastrostomy (PEG)*, *Non-Invasive Ventilation (NIV)* and death in ALS patients. This task was based on the same data and the same design as Task 1 in iDPP@CLEF 2022. The difference with respect to the previous year task is that we complemented those data with environmental data to investigate the impact of exposition to pollutants on prediction of disease progression. Since both training and test data were immediately available, we considered these submissions as position papers.

45 participants registered for iDPP@CLEF 2023 and 10 participants successfully submitted a total of 163 runs for Task 1, Task 2, and, Task 3.

Submission of participants are openly available in git repositories[5] and all the participant papers and slides are available through the iDPP@CLEF 2023 web site[6].

4 Datasets

iDPP@CLEF 2022 created a dataset, for the prediction of specific events related to ALS, consisting of fully anonymized data from 2,204 real patients from medical institutions in Turin, Italy, and Lisbon, Portugal. The dataset contains both static data about patients, e.g. age, onset date, gender, ... and event data, i.e. 18,512 ALSFRS-R questionnaires and 4,015 spyrometries.

---

[5] https://bitbucket.org/brainteaser-health/.

[6] https://brainteaser.health/open-evaluation-challenges/idpp-2023/.

iDPP@CLEF 2023 created a dataset, for the prediction of worsening of MS, consisting of fully anonymized data from 1,792 real patients from medical institutions in Pavia, Italy, and Turin, Italy. The dataset contains both static data about patients, e.g. age, gender, . . . and 2.5 years of visits. (EDSS scores, evoked potentials, relapses, MRIs).

All the datasets are highly curated and they are produced from the *Brain-Teaser Ontology (BTO)*[7] [5] which ensures the consistency of the data represented. Moreover, several checks have been performed to ensure that all the instances are clean, contain proper values in the expected ranges, and do not have contradictions.

All the datasets are available for further research on Zenodo [2] and access to them is regulated by the BRAINTEASER data sharing policy. Researchers have to submit a brief proposal, specifying the research questions, the adopted methods and algorithms, how datasets will be used, what are the expected outcomes. Before granting the access to the datasets, a committee constituted by members of the BRAINTEASER project assesses the proposal in order to ensure a proper use of the datasets and the quality and appropriateness, also from a medical point of view, of the expected claims and inferences derived from the datasets.

5   iDPP@CLEF 2024

While the previous editions focused on retrospective patient data, for ALS iDPP@CLEF 2024 will focus on prospective patient data collected via a dedicated app developed by the BRAINTEASER project and sensor data in the context of clinical trials in Turin, Pavia, Lisbon, and Madrid.

For MS, iDPP@CLEF 2024 will rely on retrospective patient data prepared in iDPP@CLEF 2023 complemented with environmental and pollution data from clinical institutions in Pavia and Turin.

In particular, we will organize the following activities:

– **Task 1 - Predicting ALSFRS-R score from sensor data (ALS)**: it will focus on predicting the ALSFRS-R score, assigned by medical doctors roughly every three months, from the sensor data collected via the app. The ALSFRS-R score is a somehow "subjective" evaluation performed by a medical doctor and this task will help in answering a currently open question in the research community, i.e. whether it could be derived from objective factors.
– **Task 2 - Predicting patient self-assessment score from sensor (ALS)**: it will focus on predicting the self-assessment score assigned by patients from the sensor data collected via the app. If the self-assessment performed by patients, more frequently than the assessment performed by medical doctors every three months or so, can be reliably predicted by sensor and app data, we can imagine a proactive application which, monitoring the sensor data, alerts the patient if an assessment is needed.

---

[7] https://w3id.org/brainteaser/ontology.

- **Task 3 - Predicting relapses from EDSS sub-scores and environmental data (MS)**: it will focus on predicting a relapse using environmental data and EDSS sub-scores. This task will allow us to assess if exposure to different pollutants is a useful variable in predicting a relapse.

We will provide prospective, fully anonymized MS and ALS clinical data including demographic and clinical characteristics as well as environmental and sensor data, coming from clinical trials currently running at institutions in Italy, Portugal, and Spain.

For Task 1 and Task 2, we will release a brand-new dataset consisting of 100 ALS patients that were followed for up to 18 months and whose progression was tracked by regular clinical evaluations. Note that even if the absolute number of patients might not seem high, this is a very rich dataset due to the daily collection of sensor data, with tens of thousands of data points in total.

For Task 3, we will re-use part of the MS dataset developed in iDPP@CLEF 2023 and we will extend it with environmental and pollution data.

6   Conclusions

iDPP@CLEF is a shared tasks focusing on predicting the temporal progression of ALS and MS and on the explainability of the AI algorithms for such prediction.

The first edition, iDPP@CLEF 2022, focused on ALS progression prediction and participation was satisfactory, hinting at the interest of the community concerning the task.

The second iteration, iDPP@CLEF 2023, we investigated MS progression prediction and how to exploit pollution and environmental data to improve progression prediction of ALS. The participation sensibly increased with respect to the previous edition, indicating the strong interest of the community on these topics.

As a further outcome of the first two iterations of iDPP@CLEF, we produced two large datasets containing retrospective clinical data about real patients and made them freely available through Zenodo.

The third iteration, iDPP@CLEF 2024, will shift the focus to prospective data about ALS patients, collected via a mobile app and dedicated sensors made available by the BRAINTEASER project, and on the prediction of ALSFRS-R scores from sensor data. For MS, we will focus on understanding the impact of pollution and environmental data on the prediction of the progression of the disease.

Acknowledgements. The work reported in this paper is partially supported by the European Union Horizon 2020 BRAINTEASER project (https://brainteaser.health/), grant agreement number GA101017598.

References

1. Cedarbaum, J.M., Stambler, N., Malta, E., Fuller, C., Hilt, D., Thurmond, B., Nakanishi, A.: The ALSFRS-R: a revised ALS functional rating scale that incorporates assessments of respiratory function. J. Neurol. Sci. **169**(1–2), 13–21 (1999)
2. Faggioli, G., et al.: BRAINTEASER ALS and MS Datasets. Zenodo, June 2023. https://doi.org/10.5281/zenodo.8083180
3. Faggioli, G., et al.: Intelligent disease progression prediction: overview of iDPP@CLEF 2023. In: Arampatzis, A., et al. (eds.) Experimental IR Meets Multilinguality, Multimodality, and Interaction. Proceedings of the Fourteenth International Conference of the CLEF Association (CLEF 2023). LNCS, vol. 14163, Springer, Heidelberg (2023)
4. Faggioli, G., et al.: Overview of iDPP@CLEF 2023: the intelligent disease progression prediction challenge. In: Aliannejadi, M., Faggioli, G., Ferro, N., Vlachos, M. (eds.) CLEF 2023 Working Notes, pp. 1123–1164, CEUR Workshop Proceedings (CEUR-WS.org), ISSN 1613–0073. https://ceur-ws.org/Vol-3497/ (2023)
5. Faggioli, G., Marchesin, S., Menotti, L., Di Nunzio, G.M., Silvello, G., Ferro, N.: The BrainTeaser Ontology. Zenodo, May 2023. https://doi.org/10.5281/zenodo.7886997
6. Guazzo, A., et al.: Intelligent Disease Progression Prediction: Overview of iDPP@CLEF 2022. In: Barrón-Cedeño, A., et al. (eds.) Experimental IR Meets Multilinguality, Multimodality, and Interaction. Proceedings of the Thirteenth International Conference of the CLEF Association (CLEF 2022), pp. 395–422. LNCS, vol. 13390, Springer, Heidelberg (2022)
7. Guazzo, A., et al.: Overview of iDPP@CLEF 2022: The Intelligent Disease Progression Prediction Challenge. In: Faggioli, G., Ferro, N., Hanbury, A., Potthast, M. (eds.) CLEF 2022 Working Notes, pp. 1130–1210, CEUR Workshop Proceedings (CEUR-WS.org), ISSN 1613–0073. http://ceur-ws.org/Vol-3180/ (2022)
8. Kurtzke, J.F.: Rating neurologic impairment in multiple sclerosis: an expanded disability status scale (EDSS). Neurology **33**(11), 1444–1452 (1983)

LongEval: Longitudinal Evaluation of Model Performance at CLEF 2024

Rabab Alkhalifa[1,2][iD], Hsuvas Borkakoty[3][iD], Romain Deveaud[4][iD],
Alaa El-Ebshihy[5,6][iD], Luis Espinosa-Anke[3,7][iD], Tobias Fink[5,6][iD],
Gabriela Gonzalez-Saez[8][iD], Petra Galuščáková[9][iD], Lorraine Goeuriot[8][iD],
David Iommi[5][iD], Maria Liakata[1,10,11][iD], Harish Tayyar Madabushi[12][iD],
Pablo Medina-Alias[12][iD], Philippe Mulhem[8(✉)][iD], Florina Piroi[5,6][iD],
Martin Popel[13][iD], Christophe Servan[4,14][iD], and Arkaitz Zubiaga[1][iD]

[1] Queen Mary University of London, London, UK
[2] Imam Abdulrahman Bin Faisal University, Dammam, Saudi Arabia
[3] Cardiff University, Cardiff, UK
[4] Qwant, Paris, France
[5] Research Studios Austria, Data Science Studio, Vienna, Austria
[6] TU Wien, Vienna, Austria
[7] AMPLYFI, Cardiff, UK
[8] Univ. Grenoble Alpes, CNRS, Grenoble INP, Institute of Engineering Univ.
Grenoble Alpes, LIG, Grenoble, France
Philippe.Mulhem@imag.fr
[9] University of Stavanger, Stavanger, Norway
[10] Alan Turing Institute, London, UK
[11] University of Warwick, Coventry, UK
[12] University of Bath, Bath, UK
[13] Charles University, Prague, Czech Republic
[14] Paris-Saclay University, CNRS, LISN, Paris, France

Abstract. This paper introduces the planned second LongEval Lab, part of the CLEF 2024 conference. The aim of the lab's two tasks is to give researchers test data for addressing temporal effectiveness persistence challenges in both information retrieval and text classification, motivated by the fact that model performance degrades as the test data becomes temporally distant from the training data. LongEval distinguishes itself from traditional IR and classification tasks by emphasizing the evaluation of models designed to mitigate performance drop over time using evolving data. The second LongEval edition will further engage the IR community and NLP researchers in addressing the crucial challenge of temporal persistence in models, exploring the factors that enable or hinder it, and identifying potential solutions along with their limitations.

Keywords: Evaluation · Temporal Persistence · Temporal Generalisability · Information Retrieval · Text Classification

1 Introduction

The second edition of the LongEval CLEF 2024 shared task continues its exploration of the temporal persistence of Information Retrieval (IR) systems and Text Classifiers, building upon, and aiming to further, the insights from the first edition [1]. We extend its focus on evaluating system performance degradation over time using evolving data, consistent with prior research [3,4,8,10,12].

The previous edition of LongEval reinforced the evidence that the performance of information retrieval and classification systems is indeed influenced by the temporal evolution of data. In this year's edition, the two tasks, retrieval and classification, are once again proposed. Task 1, related to Information Retrieval, deals with the scenario where web documents evolve over time, queries are not known in advance, relevance judgments are non-binary, and submissions must provide ranked lists as results. The Task 2 focuses on text classifiers in which the target classes are predefined while language usage associated with each class evolves rapidly over time, as in social media.

The goal of the LongEval 2014 lab is to promote the proposal of novel approaches that can automatically adapt to possible temporal dynamics in textual data. In doing so, we expect that new approaches will be able to foster time-insensitive computational retrieval and text classifiers methods. As such, the expected outcomes from this lab are threefold:

– to draw a deeper understanding of how time impacts IR and classification systems. The LongEval 2023 results need to be extended to a longer timeline to be more useful to the research community;
– to assess the effectiveness of different retrieval and classification approaches in achieving temporal persistence;
– to enable the advancement of computational methods that leverage ageing labelled datasets, while minimising performance drop over time.

The remainder of the paper is structured as follows: LongEval-Retrieval is covered in Sect. 2.1, while LongEval-Classification is covered in Sect. 2.2. Both sections present tasks and provide additional information about the data and the baselines to be used. Section 3 contains additional information and guidelines for participants.

2 Description of the Tasks

2.1 Task 1: LongEval-Retrieval

The LongEval Task 1 aims to support the development of Information Retrieval systems able to face temporal evolution. This task makes use of evolving Web data, in a way to evaluate retrieval systems longitudinally: the systems are expected to be persistent in their retrieval efficiency over time. The systems are evaluated on several collections of documents and queries, corresponding to real data acquired from a French Web search engine, Qwant[1]. The ***LongEval-Retrieval*** evaluation relies on the evaluation of **the same IR systems** on three test collections:

- Lag-3 (respectively Lag-12 and Lag-14 collection(s): a test collection acquired three (respectively twelve and fourteen) months after the last sample from the train collection

Lessons Learned from 2023. 37 teams registered for the first edition of the LongEval Retrieval task, and 14 teams submitted their runs. This number is quite large for a first edition. Several insights were learned [1]:

- No real proposal was specifically dedicated to cope with the evolution of the data
- The best approaches rely on large language model query expansion techniques
- The correlation between ranking of systems is similar for short and long lags
- The systems that are the more robust to the evolution of test collection were not the best performing ones

From these lessons, this year task enlarges the lags between the train and test collections, and provides three test collections in a way to provide a deeper understanding of the impact of the data evolution.

Data. Globally, the dataset for 2024 is twice the size of 2023, as we use the train+test data of 2023 as the 2024 train set.

1. The train dataset is composed of 4M documents (in French, translated to English), as well as 3,000 of queries with associated computed relevance assessments from a simplified Dynamic Bayesian Network (sDBN) Click Model [5,6] acquired from real users of the French Qwant search engine. We will require the participants to not make any use of the assessments provided on the 2023 collections, but only the data of the 2023 train set.
2. The test collection is composed of three sub-collections: Lag-3, Lag-12 and Lag-14, corresponding to data acquired at several lags (3, 12 and 14 months from the last train data). Each of these test collections is similar to the train set, except that they do no contain any relevance assessments. The participants are expected to submit runs for each of the three lag collections, using the same system, i.e. a system trained only on the train dataset.

The total data for this task will be composed of 8 million documents and 6,000 queries, provided by Qwant[2]. Each document set will have a release time stamp, with the first set (in chronological order) being the training data.

---

[1] Qwant being mostly used by French speaker, it explains why it is easier to gather data (user queries and documents) in this language rather than English.
[2] Qwant search engine: https://www.qwant.com/.

Evaluation. The submitted systems will be evaluated in two ways:

1. **nDCG** scores calculated on each lag test set provided for the sub-tasks. Such a classical evaluation measure is consistent with Web search, for which the discount emphasises the ordering of the top results.
2. **Relative nDCG Drop (RnD)** measured by computing the difference between nDCG values between Lag-3 and Lag-12 datasets, Lag-3 and Lag-14 datasets as well as between Lag-12 and Lag-14 datasets. Such values will allow to check the robustness of systems against the evolution of the data.

These measures assess the extent to which systems provide good results, but also the extent to which they are robust against the data (queries/documents) evolution along time. Using these evaluation measures, a system that has good results using nDCG, and also good results according to the RnD measure is considered to be able to cope with the evolution over time of the Information Retrieval collection.

2.2 Task 2: LongEval-Classification

Detecting the stance in social media posts is essential [9,11]. Yet, comprehending the evolution of social media stances over time poses a significant challenge [2,4], a topic that has gained recent interest in the AI and NLP communities but remains relatively unexplored. The performance of social media stance classifiers is intricately linked to temporal shifts in language and evolving societal attitudes toward the subject matter. In LongEval 2024, social media stance detection, a multi-label English classification task, takes center stage, surpassing the complexity of the binary sentiment task in LongEval 2023 [1]. Its primary aim is to assess the persistence of stance detection models in the dynamic landscape of social media posts.

The *LongEval-Classification* organizes two sub-tasks.

   Sub-task 2.A: Short-term persistence. In this sub-task participants will develop models which demonstrate performance persistence over short periods of time, i.e. using test set within 2–3 years apart from the training data.

   Sub-task 2.B: Long-term persistence. In this sub-task participants will develop models which demonstrate performance persistence over longer period of time, i.e. test set within 4–5 years apart from the training data and also distant from the short-term test set.

   The insights from the first edition of LongEval shed light on text classifiers' performance drop and highlighted it importance to the research community. To boost engagement and participation, we will release the collaboration (colab) platform, data, and starting kit ahead of time. This streamlined access encourages wider involvement from the research community. In the 2024 edition, we have expanded the dataset, focusing on a three-way classification task for stance detection. This expansion enriches research opportunities and addresses nuanced challenges in temporal persistence for text classification and opinion dynamics. The evolving nature of the task continues to uncover new dimensions in understanding temporal persistence and adaptability of text classifiers over time.

Data. In this task, we will make use of, and extend with new annotations, the Climate Change Twitter dataset [7]. Our primary focus will be on climate change stance, time of the post (created at), and the textual content of the tweets, which we will refer to as the **CC-SD** dataset. This **CC-SD** is large-scale, covering a span of 13 years and containing a diverse set of more than 15 million tweets from various years. Using the BERT model to annotated tweets, the **CC-SD** stance labels in three categories: those that express support for the belief in man-made climate change (believer), those that dispute it (denier), and those that remain neutral on the topic. The total sum of the categorized tweets over all time span are as follows: 11,292,424 tweets as believers, 1,191,386 as deniers, and 3,305,601 as neutral, distributed across the timeline. The annotation is performed using transfer learning with BERT as distant supervision based on another sentiment climate change dataset[3] and, thus, can be easily manually annotated to improve its precision using manual annotation. We plan to release data in two phases:

1. In the **practice phase**, participants will be given **(1) a distantly anno-tated training set sampled from CC-SD** (tweet, label) created over a time interval t. Such data is dedicated for model training, as well as **(2) human-annotated "within time" practice set** (tweet, label) from the same time period t. **(3) human-annotated "short time" practice set** (tweet, label) distant from time period t. These two human practice sets are intended to allow participants to develop their systems before the following evaluation phase, and will not be used to rank their submissions. All these resources, including python-based baseline code and evaluation scripts, will be made available to participating teams upon data release.

2. In the **evaluation phase**, participants will be provided with three human-annotated testing sets without their labels (id, tweet): **(1) "within time"** acquired during time period t, **(2) short-term** acquired during a time interval t' occurring shortly after t (with no intersection between t and t') dedicated to evaluate *short-term persistence (sub-task 2.A)*, and **(3) long-term** acquired long after t during a time interval t'' (with no intersection between t and t'') dedicated to evaluate *long-term persistence (sub-task 2.B)*. Similarly to Task 1, participating teams are required to provide a per-formance score for the "within time" test set, even if they are interested in one of the sub-tasks to calculate persistence metrics, i.e. RPD.

Evaluation Metrics. Evaluation metrics for this edition of the task remain con-sistent with the previous version. All submissions will be assessed using two key metrics: the **macro-averaged F1-score** on the corresponding sub-task's testing set and the **Relative Performance Drop (RPD)**, calculated by comparing performance on "within time" data against results from short- or long-term dis-tant testing sets. Submissions for each sub-task will be ranked primarily based on the macro-averaged F1-score. Additionally, a unified score, **the weighted-F1,**

---

[3] https://www.kaggle.com/datasets/edqian/twitter-climate-change-sentiment-dataset.

will be computed between the two sub-tasks, encouraging participants to contribute to both for accurate placement on a collective leaderboard and a deeper analysis of their system's performance in various settings.

Baseline. Participants are expected to propose temporally persistent classifiers based on state-of-the-art computational methods. The goal is to achieve high weighted-F1 performance across short and long temporally distant test sets while maintaining a reasonable RPD when compared to a test set from the same time period as training. We intend to use **BERT**[4] [7] as a baseline classifier.

3   LongEval Timeline

Information and updates about the LongEval Lab, and the submission guidelines, will be communicated mainly through the lab's website[5]. The training data for both tasks will be released in December 2023, and the test data in February 2024. Participant submission deadline is planned for May 2024, with the evaluation results to be released in June 2024. During the CLEF 2024 conference, LongEval will organize a workshop, with participant presentations as well as invited speakers. The workshop will welcome other submissions on the topic of temporal persistence that were not part of the shared task.

Acknowledgements. This work is supported by the ANR Kodicare bi-lateral project, grant ANR-19-CE23-0029 of the French Agence Nationale de la Recherche, and by the Austrian Science Fund (FWF, grant I4471-N). This work is also supported by a UKRI/EPSRC Turing AI Fellowship to Maria Liakata (grant no. EP/V030302/1). This work has been using services provided by the LINDAT/CLARIAH-CZ Research Infrastructure (https://lindat.cz), supported by the Ministry of Education, Youth and Sports of the Czech Republic (Project No. LM2023062) and has been also supported by the Ministry of Education, Youth and Sports of the Czech Republic, Project No. LM2023062 LINDAT/CLARIAH-CZ.

References

1. Alkhalifa, R., et al.: Extended overview of the CLEF-2023 longeval lab on longitudinal evaluation of model performance (2023). https://api.semanticscholar.org/CorpusID:259953335
2. Alkhalifa, R., Kochkina, E., Zubiaga, A.: Opinions are made to be changed: temporally adaptive stance classification. In: Proceedings of the 2021 Workshop on Open Challenges in Online Social Networks, pp. 27–32 (2021)
3. Alkhalifa, R., Kochkina, E., Zubiaga, A.: Building for tomorrow: assessing the temporal persistence of text classifiers. Inf. Process. Manag. **60**(2), 103200 (2023)
4. Alkhalifa, R., Zubiaga, A.: Capturing stance dynamics in social media: open challenges and research directions. Int. J. Digital Humanit. 1–21 (2022)

---

[4] https://huggingface.co/bert-base-uncased.
[5] https://clef-longeval.github.io.

5. Chapelle, O., Zhang, Y.: A dynamic bayesian network click model for web search ranking. In: Proceedings of the 18th International Conference on World Wide Web, pp. 1–10. WWW 2009, Association for Computing Machinery, New York, NY, USA (2009). https://doi.org/10.1145/1526709.1526711

6. Chuklin, A., Markov, I., Rijke, M.D.: Click models for web search. Synth. Lect. Inf. Concepts Retrieval Serv. **7**(3), 1–115 (2015). https://doi.org/10.2200/S00654ED1V01Y201507ICR043

7. Effrosynidis, D., Karasakalidis, A.I., Sylaios, G., Arampatzis, A.: The climate change twitter dataset. Expert Syst. Appl. **204**, 117541 (2022). https://doi.org/10.1016/j.eswa.2022.117541, https://www.sciencedirect.com/science/article/pii/S0957417422008624

8. Florio, K., Basile, V., Polignano, M., Basile, P., Patti, V.: Time of your hate: the challenge of time in hate speech detection on social media. Appl. Sci. **10**(12), 4180 (2020)

9. Küçük, D., Can, F.: Stance detection: a survey. ACM Comput. Surv. **53**(1), 1–37 (2020). https://doi.org/10.1145/3369026

10. Lukes, J., Søgaard, A.: Sentiment analysis under temporal shift. In: Proceedings of the 9th workshop on Computational Approaches to Subjectivity, Sentiment and Social Media Analysis, pp. 65–71 (2018)

11. Mohammad, S.M., Sobhani, P., Kiritchenko, S.: Stance and sentiment in tweets. ACM Trans. Internet Technol. **17**(3), 1–23 (2017). https://doi.org/10.1145/3003433, http://alt.qcri.org/semeval2016/task6/

12. Ren, R., et al.: A thorough examination on zero-shot dense retrieval (2022). arxiv:2204.12755). https://doi.org/10.48550/ARXIV.2204.12755, https://arxiv.org/abs/2204.12755

Author Index

N. Goharian et al. (Eds.): ECIR 2024, LNCS 14613, pp. 67–68, 2024.
https://doi.org/10.1007/978-3-031-56072-9